Narrow Gauge Rails
Bosnia-Hercegovina

Keith Chester

Bosn.-herc. Staatsbahnen.

This book is dedicated to Michael Schumann and to the memory of Geoffrey Smith

Readers interested in finding out more about the narrow gauge railways of Bosnia-Hercegovina are referred to the following full-length books by Keith Chester:

The Narrow Gauge Railways of Bosnia-Hercegovina Malmö 2006; 2nd edition 2008 (pub. Stenvalls)
Bosnia-Hercegovina: Narrow Gauge Album Malmö 2010 (pub. Stenvalls)

Further, recently published articles by the author on specific aspects of these fascinating railways include:

'Albori' *Industrial Railway Record 207*. [Melton Mowbray] December 2011 pp. 246–247
'The Genesis of the Ugljevik–Bosanska Rača Coal Railway in Bosnia-Hercegovina' *The Industrial Locomotive 143*. [Swindon] June 2012 pp. 84–94
'"Foreign" Locomotives on Loan to the Bosnisch-Herzegowinische Landesbahnen during WW1 – Part I' *Locomotives International 88*. [Chippenham] February 2014 pp. 49–56
'"Foreign" Locomotives on Loan to the Bosnisch-Herzegowinische Landesbahnen during WW1 – Part II' *Locomotives International 89*. [Chippenham] April 2014 pp. 48–52
'Assassination in Sarajevo' *Narrow Gauge World 94*. [Southend-on-Sea] June 2014 pp. 35–38
'The Great Flood of 1915' *Narrow Gauge World 100*. [Southend-on-Sea] January 2015 pp. 24–26
'"On the road to nowhere… "' – The k.u.k. Militärbahn Simin Han – Bulatovci' in Wallisch-Pertl C. (ed.) *Selection* Vienna 2014 pp. 76–87 (pub. bahnmedien.at)

Published by Mainline & Maritime Ltd
3 Broadleaze, Upper Seagry, near Chippenham, SN15 5EY
Tel: 01275 845012
www.mainlineandmaritime.co.uk
orders@mainlineandmaritime.co.uk
Trade and Retail Enquiries welcome!
Printed in the UK
ISBN: 978-1-900340-39-7

Front Cover
A smokey 2-8-0 no. 84-002 (Alco 55098/1915) shunts at Sarajevo in July 1966.
Peter Gray

Rear Cover (upper)
The driver of 0-6+4R no. 97-004 (WLF 958/1895) takes advantage of a brief halt at Oborci to inspect his locomotive.
Peter Gray

Rear Cover (lower)
2-8-2 no. 85-035 (Bp 5064/1931) and an unidentified class 83 0-8-2 catch the last rays of the evening sun near Ustiprača at the head of an eastbound freight.
John Millbank

Title Page
Klose radial tank no. 189-008 was constructed by Krauss, Linz in 1889, works number 2185.
Courtesy Željeznički Muzej, Belgrade

FOREWORD

It was at an autumn meeting of the Continental Railway Circle in London in 1960 I made my decision to visit Yugoslavia the following summer to explore the narrow gauge railway network. Dusty Durrant and Trevor Rowe had already made brief visits to that country and were at the time our sole sources of information as to what was to be found there. To the narrow gauge enthusiast the country appeared to have great potential. There was a metre gauge system in the north of the country close to the Hungarian border and in the far south-east in Macedonia a lengthy two-foot gauge railway. However, the main attraction was the world's most comprehensive network of 760mm narrow gauge railways, which served Bosnia, Hercegovina, and Serbia, although it was known that this was progressively being converted to standard gauge. Apart from the passenger carrying railways it was also known that in Bosnia-Hercegovina there were many forestry railways but the details of these were hard to come by.

So it was that in the summer of 1961 three students set out in an elderly Land Rover to explore. The Yugoslav road network was 99% unsurfaced, so progress was slow. Most of the trains we saw were full to bursting point and some had covered wagons being used to carry passengers. Most of the locations we visited were outside the tourist areas where foreigners were viewed with curiosity or suspicion. We met hospitable people and drank Slivovitz, and also policemen to whom a camera was like a red rag to a bull. Evidently, the people had been told that they had been invaded so quickly at the start of the Second World War because German tourists had photographed all important buildings and structures in the years beforehand. We were arrested on this basis on the first occasion in Osijek because we might have taken photographs at the railway station. We followed the policeman on his bicycle to the police station where once they had found an interpreter everything was sorted out. The pattern was set for railway hunting in Yugoslavia where on average the police arrested us once a week.

This first visit to Yugoslavia had me hooked, and many more visits followed. You can look at railways simply because they exist, but after a while you ask why they exist. My university library answered many of my questions and in doing so created more. Later, the British Museum and the University of London Department of Slavonic Studies libraries provided a lot more information.

Various people I heard were collecting information for a book on the history of the narrow gauge railways of Yugoslavia but none of these books came to fruition. Further I realised that I was not a natural author. Thus when in the mid-1990s Keith Chester first approached me asking for support in fulfilling a commission he had received to write a short history of the Yugoslav narrow gauge railways I felt a sense of great relief. I suggested to him that the subject deserved a comprehensive history including the social and political background to their construction. The speed with which he warmed to this idea suggested that he already had this in mind.

It is therefore with great pleasure that I write this foreword to Keith's third and latest book on the subject. He writes in a highly readable way and he is very thorough in his research gaining much support from fellow enthusiasts. He has performed us all a great service and his writing has brought to light many new sources of information. May this process continue.

Michael Schumann

75-year-old no. 189-004 (KrLi 1797/1886) runs light engine near Doboj Novi on 11 August 1961. One of the photographer's many arrests in Yugoslavia followed shortly afterwards.

Michael Schumann

The first railway in Bosnia-Hercegovina was the approximately 100km-long standard gauge line between Banja Luka and Dobrljin opened in late 1872 by the CO, the northernmost section of a grand trunk route intended to connect Istanbul, the capital of the Ottoman Empire, with Central Europe. No. 401 (Tub 122/1872) was one of three 0-4-0Ts found in Banja Luka at the time of the Austro-Hungarian occupation in 1878. By the mid-1890s the loco was in Austria, where its history is surprisingly vague. It is known to have worked briefly on the Steinfeld system, which served munitions factories in Lower Austria; it was also employed on the military training railway at Korneuburg northeast of Vienna where it is seen testing a wooden bridge newly constructed by pioneers.

Keith Chester collection

For the first three or four years after it opened the kkBB relied on its fleet of 24 small contractors' 0-4-0Ts to operate its 190km-long main line between Bosanski Brod and Zenica. It was only with the introduction of first the 'Duplex' 0-4-0T+0-4-0Ts (1881) and then the Klose radial 0-6+2Ts (1885) that they began to be replaced: most were sold off, but a few were retained for light shunting duties or working engineering trains. The now preserved *Rama* (KrMü 264/1873) and *Ida* (KrMü 404/1874) were employed on the Ostbahn contract in 1904–06. However, with no further railway construction likely, in 1908 *Ida* was sold to Johann Bleckmann for his Phönix steelworks at Mürzzuschlag in Austria where it worked until c.1931.

Alfred Moser collection

KrMü 751/1878 remained on the books of the kkBB and BHStB longer than most of the Hügel & Sager 0-4-0Ts. This 40hp contractors' loco had its fifteen minutes of fame when it was displayed (alongside 2-4-2 no. 105 and 0-6+2T no. 230) at the 1896 Millennium Exhibition in Budapest. It was then moved to Austria for the 1898–99 Ybbstalbahn contract where it is seen at the small station at Gaming in 1898.

Ludwig Fürnweger collection

Its fate after this contract is unknown until August 1907 when it was acquired by Ács sugar factory in Hungary and given the name *Lili*. Looking well cared, it was photographed at Ács in the 1940s. As GVI no. 292,044 withdrawal took place on 20 September 1956.

Tibor Nagy collection

Brod n. S. Most.

In 1878 Bosanski Brod was chosen as the location of the northern terminus of the 760mm gauge military field railway being built into the Bosnian interior because it offered plenty of space for the transhipment facilities that would be needed. The following summer a wooden bridge was completed across the river Sava, extending the standard gauge MÁV line from Slavonski Brod to Bosanski Brod. Ever increasing levels of traffic compelled the successors of the field railway, the kkBB, the BHStB and BHLB, to undertake the Sisyphean task of expanding the area and equipment of the interchange at regular intervals. Sometimes, temporary tracks were laid at Bosanski Brod to increase capacity. The viaduct in the background was erected in 1882–84 to replace the temporary structure of 1879.

Keith Chester collection

The railway was extended south from Zenica to Sarajevo in 1881–82. It was initially worked by pairs of 0-4-0Ts coupled back-to-back but this arrangement proved unsatisfactory. As soon as possible the 'Duplexes' were split and the single locos used on shunting and other light duties. In the mid-1890s an unidentified 'half-duplex' stands outside the central repair shops in Sarajevo.

Keith Chester collection

No. 19 (KrLi 1161/1882) is decked out in bunting for the 1910 May Day celebrations. By the time of this photograph, these 0-4-0Ts had outlived their usefulness and were scheduled for withdrawal in the coming decade. The outbreak of the Great War prevented this and half-a-dozen were still in stock in the late 1930s.

Alfred Moser collection

No. 189-016 (KrLi 2691/1892) brews up at Jajce on 20 August 1952. Spindly, almost delicate though this six-coupled tank may look to modern eyes, it was nonetheless the game changer on the 760mm gauge railways in Bosnia-Hercegovina. The application of the Klose system of articulation permitted the operation of engines with longer wheelbases, and hence engines with larger and more powerful boilers. Faster and heavier trains could now be run and for nearly two decades Klose locomotives ruled the roost on the kkBB and BHStB.

J.C. Gillham / A. Peters collection

Further development of the Klose type came to an abrupt halt in 1903 with the introduction of a conventional 0-8-2, which took full advantage of Gölsdorf's recent innovations with sideplay. No. 1002 (KrLi 5069/1903) was the second of the initial 29, which were 350hp compounds; from 1909 onwards, however, they were delivered as 400hp superheated simples. In all 185 of these remarkably successful locos were built between 1903 and 1949. At the time of writing (2015) one is still steamed regularly at Banovići.

Lothar Rihosek collection

Gruss aus Travnik   Pozdrav iz Travnik-a, 21./II. 1904

281. Rosner & Laufer, Wien

The station at Travnik at the turn of the century. Once the capital of Ottoman Bosnia and Hercegovina, Travnik was the most important town on the railway from Lašva to Bugojno and Jajce opened in stages between 1893 and 1895. This was the first leg of a scheme to link the industries of central Bosnia with the seaport of Split (via Bugojno) and with the standard gauge k.u.k. Militärbahn at Banja Luka (via Jajce). There were strong economic arguments for the two lines. However, a thriving port at Split in the Austrian province of Dalmatia would have rivalled Hungary's great port of Rijeka (in its province of Croatia).

Keith Chester collection

Plivasbrücke, Station Jajce.

On its approach to Jajce from the south, the railway crossed the river Pliva on a substantial viaduct before veering left to the station. The proposed Jajce–Banja Luka line would, in conjunction with the putative Spalato-Bahn, have created a rail link between Austria and Dalmatia. Moreover, it would have broken Hungary's monopoly over all railway traffic to Bosnia-Hercegovina's narrow gauge system by virtue of the MÁV operating the sole railway into Bosanski Brod. Given the fact that the province took 70% of its imports from Austria and only 30% from Hungary, this was a major consideration for Vienna. The Hungarians ('the enemy within' jibed the heir apparent Franz Ferdinand) opposed both the Spalato-Bahn and the Banja Luka–Jajce link implacably. The two railways joined the long list of commercial and political disputes between Vienna and Budapest; and were never built.

Keith Chester collection

This photograph of 0-6+2 rack and adhesion tank no. 602 (WLF 740/1890) taken at Dolac MPD in 1911 has suffered from the ravages of time but is one of the very few we have of any of these engines in service. With only 250hp at their disposal, they soon proved underpowered for the rapidly growing traffic on the 35–60% (1 in 28.57–16.67) grades over the Ivan pass, and were not duplicated. Instead, an enlarged 350hp tender version was introduced in 1894. As more of these 0-6+4Rs became available (they eventually totalled 38), so the original eight 0-6+2RTs were all gradually transferred to the less challenging rack line over the Komar pass where they remained until they were withdrawn as JDŽ class 195 in the late 1930s.

Courtesy Železnički Muzej, Belgrade

No. 501 (KrLi 2859/1893) stands at Sarajevo works in the latter half of the 1890s. Renumbered from BHStB no. 61 in 1895, the Klose 0-10+2T was a bold experiment to investigate whether the rack-fitted Komar pass between Travnik and Donji Vakuf could be worked by an adhesion locomotive. Trials suggested it was powerful enough, but this narrow gauge behemoth badly damaged the track, and was quietly set aside. Subsequently, attempts were made to employ it on the newly opened Ostbahn, which had been laid with heavier rails, but derailments were frequent and in January 1908 its use on passenger trains was barred. Thereafter it led something of a shadowy existence, eking out meagre mileages over the next three decades on steam-heating, shunting and banking duties.

Courtesy Železnički Muzej, Belgrade

In her 1909 book *Through Bosnia and Herzegovina With a Paint Brush*, Mrs E.R. Whitwell, an evidently rather prim English middle-class lady, describes travelling by train through 'mountains of rugged grey stone. The railway goes up a great height and winds in and out of the hills, the view is more grand than beautiful. The train stops at many little stations for the sole reason of allowing the travellers to have drinks, we concluded, as at many places we stayed for ten minutes for apparently no reason except we saw the travellers flocking into a bar.' She added 'The journeys in these mountainous regions are very long and the trains creep very slowly.'

Michael Schumann collection

In 1901 a passenger boards one of Sarajevo's 760mm gauge trams standing in front of the main railway station. The circular route indicator shows that it will be traversing the full three kilometres back to the town centre. No. 15 was the last of a batch of five 20hp tramcars delivered by Weitzer of Graz for the 1895 electrification of the horse-worked tramway, which had opened on 1 January 1885. The new electric trams proved very popular and in 1897 two more were added to the fleet. 13 larger cars were acquired between 1908 and 1914, allowing the withdrawal of the original seven, though some of these were rebuilt as trailers (at a cost of K4,865 each) and gave another decade of service.

Courtesy Zemaljski Muzej, Sarajevo

A stylish piece of Siemens-Schukertwerke publicity for steeple cab no. 11. Delivered in 1903, this was the third electric locomotive acquired by the Sarajevo tramway to work the not inconsiderable freight and mail traffic over the tramway between the BHLB terminus and the Stadt-Bahnhof in the centre of the Bosnian capital: horses and steam engines both proving unsuitable on such workings through the streets of an expanding and modernising conurbation. Siemens supplied the electrical parts and Weitzer of Graz was responsible for the bodywork. This 60hp loco remained in service into the 1950s.

Alfred Moser collection

Even before the opening of the Ostbahn in July 1906 the sweeping and rather elegant viaduct at Bistrik was a favoured subject for commercial postcards, such as this one dated 19 September 1905. Track laying and ballasting as far as Bistrik was completed over the summer of 1904, with the first trains over the viaduct hauled by 'half-duplex' no. 26 (KrLi 1438/1883). The old centre of Sarajevo is visible between the pillars and Bistrik was the preferred station for joining and leaving trains on the Ostbahn for many of the city's inhabitants.

Keith Chester collection

Originally mooted in the late 1890s as the first leg of a railway between Sarajevo and Mitrovica, thereby providing the Monarchy with a direct railway link with Thessalonica and the Orient, the strategic and hugely expensive Bosnian Ostbahn was from the very beginning an enormously contentious issue. Construction began in 1902 and, though only 166.5km long, it took four years to complete. Much of the line had to be hacked out of rock, such as, upper left, near Međeđa, an image sent as a New Year's greeting card on 30 December 1903.

Work has progressed somewhat further in this postcard of the section between Dobrun and Vardište

Both: Keith Chester collection

Five 'half-duplex' 0-4-0Ts (nos. 11, 18, 24, 25 and 26) have been identified as working on the Ostbahn contract, one of which is seen here working a ballast train in the latter stages of construction. The first revenue earning trains on the Ostbahn were three military transports between Ustiprača and Sarajevo worked by 0-6+2T no. 209 (KrLi 2209/1890) between 5 and 7 May 1906.

Courtesy Zemaljski Muzej, Sarajevo

On 1 August 1906 the first train between Dobrun and Vardište halts in mid-section for the photographer. The occasion is decidedly low key. The same had been the case almost a month previously for the inauguration of the main line of the Ostbahn from Sarajevo to Uvac, together with the branch from Most na Limu as far as Dobrun. Unlike the official openings of other railways in Bosnia-Hercegovina, that of the Ostbahn was conducted with little fanfare. Given the international furore which had greeted the news of its authorisation in 1900, this was politically expedient. On 31 May 1906 István Burían, the minister with responsibility for Bosnia-Hercegovina, curtly informed the Landesregierung in Sarajevo that obligations in Vienna would prevent him, and other high-ranking officials, from attending any opening ceremony. Sarajevo was instructed to 'simply begin the timetabled service on 1 July'. In the event this took place on 4 July.

Courtesy Zemaljski Muzej, Sarajevo

A Klose 0-6+2T runs round its train at what looks to be the very recently opened terminus of the Ostbahn at Uvac. This was a small village (pop. 230 in 1918) hard on the Turkish frontier and was provided with generous facilities as it was expected to be the border station between the Habsburg and Ottoman Empires when the railway was eventually extended south through the Sanjak of Novibazar to Mitrovica. However, when the construction of this line was officially proposed in January 1908, it caused outrage throughout Europe, and contributed significantly to the growing distrust between Vienna and Belgrade which in July 1914 would spill over into the Great War.

Courtesy Zemaljski Muzej, Sarajevo

Sarajevo-Višegrad – Limmündung

A couple of kilometres east of Međeđa, the Ostbahn crossed the river Drina on a 130m-long viaduct at the point where it was joined by the river Lim. The railway then split into two, with the main line to Uvac following the Lim valley and the branch to Vardište continuing along the banks of the Drina as far as Višegrad. Between the opening of the railway in 1906 and its closure in 1978, history would dictate that there would be no fewer than four different bridges spanning the river.

Above: The original structure, the most graceful of them all, was probably blown up by retreating Habsburg forces as the Serbs advanced on Sarajevo in September 1914: the date of its destruction does not appear in the extensive surviving documentation on its rebuilding.

Keith Chester collection

Left: Following the recapture of the lost territory in southeast Bosnia in November 1915, Austro-Hungarian sappers set about the restoration of the bridge using prefabricated Roth-Waagner modules, a task completed on 24 April 1916. The twisted girders of the destroyed viaduct lie forlornly in the river.

Keith Chester collection

The temporary viaduct was replaced by a permanent one in 1924–25 by the Louis Eilers Stahlbau GmbH & Co. of Hannover, a German reparations payment. As the work nears completion, a pair of Henschel 2-6-6-0s (some of which came from yet another tranche of German reparations) test the new bridge. It would be destroyed by the Partisans in November 1944 in an attempt to hinder the German withdrawal from the Balkans.

Keith Chester collection

A class 85 2-8-2 trundles over the rather mundane post-1945 viaduct. The colour photograph, taken in the autumn of 1972, makes very clear the confluence of the rivers Drina and Lim. Following the closure of the railway in 1978, the bridge was briefly used for road traffic before it was flooded by a hydro-electric scheme in the 1980s.

Peter Lemmey

With over half a century of service behind it, a rather grubby no. 73-015 (KrLi 6481/1911) awaits its next turn at Lašva on 24 August 1961.

Trevor Rowe

The most elegant of all the locomotives to have graced the tracks of the narrow gauge railways of Bosnia-Hercegovina were undoubtedly the IIIb 5 (JDŽ class 73) 2-6-2s, 23 of which were put into service between 1907 and 1913. In a lengthy memorandum, dated 20 June 1906 and seeking authorisation for these engines, BHStB Director Schnack sang their praises, noting that the type, still only known by its Krauss project number L 200 b, was 'approximately 50% more efficient than the passenger locomotives currently in service', the Klose 2-4-2s of 1894 (JDŽ class 178). Moreover, if something was to be done about 'the justified complaints of passengers about overcrowded trains, then it is a matter of some urgency to acquire more powerful locomotives for the Doboj–Sarajevo section'. Assuming Krauss-Linz could 'guarantee that the new locomotive L 200 b could in all conditions haul loads of 150t at 40km/h up a grade of 8% [1 in 125]', the intention was to run trials with one loco for six months, and then acquire seven more.

Perhaps what is most revealing about this document is what Schnack planned to do with the new locos, incidentally the first on the BHStB with superheat, and how this would affect motive power dispositions on the railway in general. 'The Direktion intends to use these more powerful locos only on the Doboj–Sarajevo line, where today eight passenger engines of series IIa 4 [JDŽ class 178] are in service. Passenger trains between Doboj and Bosanski Brod should subsequently be worked by locomotives of series IIIa 4 which will be all the more easier when the current passenger locos [the 2-4-2s] are used partly on the Konjic–Mostar–Metković line and partly on the Doboj–Donja Tuzla line. In this way freight locomotives of classes IIIa 4 [Klose 0-6+2T, JDŽ class 189] and IIIa 5 [Klose 0-6+4, JDŽ class 185] at present working these trains would be freed up. The locos of class IIIa 4, which are tank engines, could then be employed on shunting turns, replacing the "half-duplexes", which have for long been inadequate to this task. The newly available locos of class IIIa 5 would be a welcome addition to the fleet of freight locomotives.'

The impact of the Prairies was immediate. In 1910, by which time 12 were in service, they ran an average of 49,667km each (compared to the annual average for all BHLB locos of 36,396km); the figures for the other modern conventional locos, the 0-8-2s, were 40,636 and 43,156km respectively for the compound and superheated variants. Moving the Klose 2-4-2s to the flatter sections of the Narentabahn made sense and these rather petite engines put in many years of service there, often double-headed, until it was upgraded after the Great War and larger locos took over the express turns.

As Schnack's memo suggests, most passenger trains out of Brod remained in the hands of the light weight radial tanks, again until the 1920s. Increasing the capacity of the Brod to Sarajevo main line was a vital issue at a time of rising traffic. Work had begun on laying heavier rails north of Sarajevo in 1896 but this proceeded at almost snail's pace. It was not until 1912 that Doboj (km 185) was reached, finally permitting the use of the powerful 0-8-2s with their 8t axle loadings. It speaks volumes of the parlous state of the finances of the Monarchy and, in particular, of the interminable bickering and politicking between Vienna and Budapest that this strategic main line was not relaid earlier and not provided with an adequate fleet of the proven 0-8-2s to work it.

Austria-Hungary had in general a poor record regarding the development of the agricultural potential of Bosnia-Hercegovina. An exception to this was the growing and processing of tobacco. Among the earliest legislation passed after the occupation was that of September 1879 setting up state salt and tobacco monopolies. Tobacco cultivation increased rapidly and it became an important cash crop, particularly in Hercegovina. Factories were established at Sarajevo, Banja Luka, Mostar and Travnik; these were manufacturing about 100,000,000 cigarettes annually by the Great War. Tobacco provided employment for large numbers of people and was a lucrative source of income for the state.

In the Neretva valley there was a large depot at Čapljina where raw tobacco was collected. Like the four factories, this was rail connected. Erected in 1881, the largest tobacco factory was at Sarajevo, where a siding led off the tramway at Marijin dvor. As this card, posted in September 1918, shows, the internal sidings were not electrified and presumably shunting was manual or by horse. On 2 May 1906 workers at the factory went on strike for higher wages. Arrests made the following morning provoked a large demonstration in front of the City Hall, during which five protesters were shot dead by gendarmes, sparking a wave of industrial unrest, the worst experienced in Habsburg Bosnia-Hercegovina. This only ended after more deaths and, finally, recognition by the authorities that a more conciliatory approach to the interests of the workers was necessary.

The tobacco factory at Travnik was connected to the railway station with a 700m-long siding; the Lašva–Donji Vakuf main line passes in the foreground.

All: Keith Chester collection

Чапљина.
Čapljina.

Gruss aus { Sarajevo.
Pozdrav iz {

Tabak-Fabrik

Verlag P. Krisč, Sarajevo.

Travnik   Duhanska tvornica — Tabakfabrik

Direktion der bosnisch=hercegovinischen Landesbahnen.

The announcement of Austria-Hungary's decision to annex Bosnia-Hercegovina on 6 October 1908 almost brought Europe to war over the winter of 1908–09. However, by removing some of the anomalies surrounding Habsburg rule in the province, it strengthened the Monarchy's position there. One immediate consequence was the renaming of the Bosnisch-Herzegowinische Staatsbahnen (BHStB) as the Bosnisch-Herzegowinische Landesbahnen (BHLB) on 12 December 1908. It took a while for the stationery to catch up.

In the first three decades of the occupation the population of Sarajevo swelled from 21,377 in 1879 to 51,919 in 1910. Much of the increase came from the influx of a burgeoning army of civil servants, contractors etc. (and their families) from outside the province: in the 1910 census one in three of Sarajevo's inhabitants claimed the citizenship of either Austria or Hungary. Members of Bosnia-Hercegovina's new élite, buffed up in their Sunday best, pose for their photograph in six-wheel saloon no. As 81 attached to a lengthy passenger train awaiting departure in Sarajevo station c.1906.

Zemaljski Muzej, Sarajevo

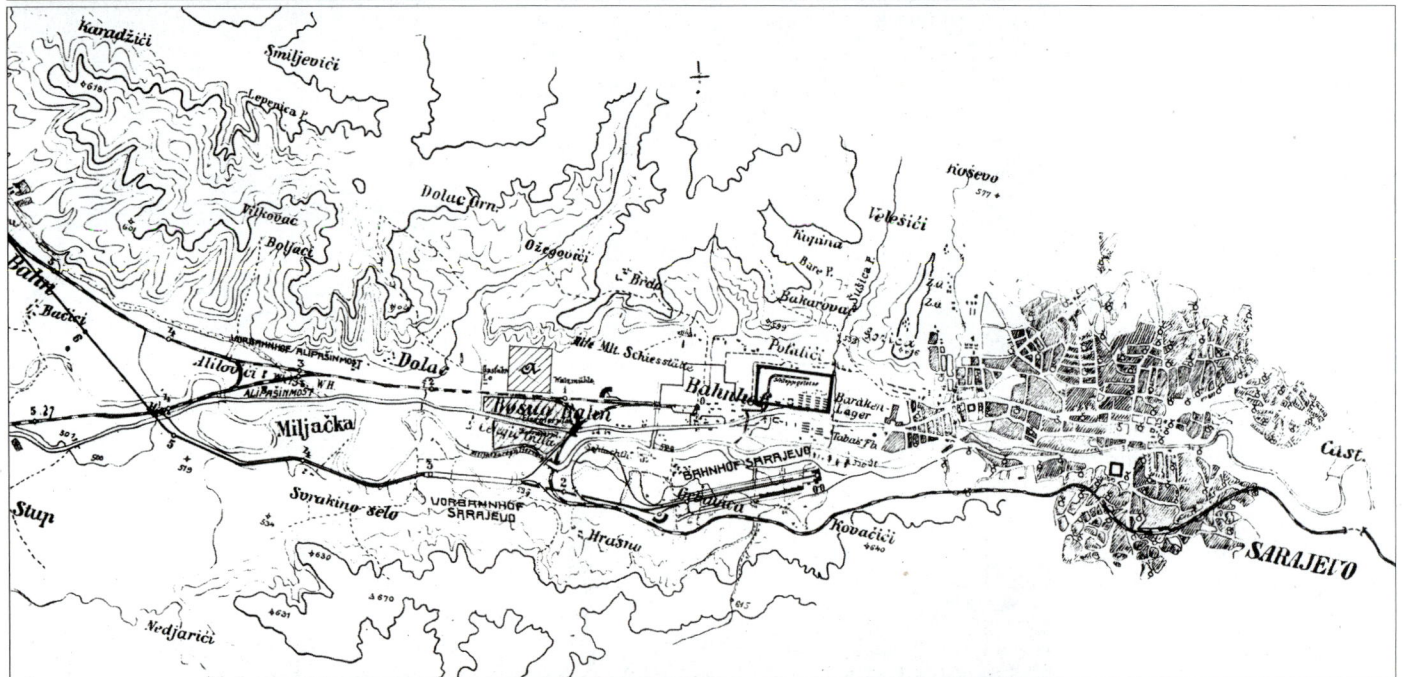

During the first decade of the twentieth century the limitations of the narrow gauge in Bosnia-Hercegovina became increasingly apparent. Rapidly rising levels of traffic (gross tonnages per km rose from 156 million in 1896 to 710m in 1910, much of it concentrated on the Bosanski Brod–Sarajevo main line) led to calls to regauge at least parts of the system. The deteriorating situation in the Balkans gave the military considerable leverage over the politicians in both Vienna and Budapest, and in 1912–13, against the background of the Balkan Wars, a compromise was finally hammered out. At the heart of this lay a 1435mm gauge railway from Šamac to Sarajevo. The map shows the planned new standard gauge station in the Bosnian capital. Having shared the same alignment on the approach to Sarajevo, the narrow and standard gauges diverge at km 5.5, with the latter running to a completely new terminus some 1.5km closer to the town centre. This would eventually have entailed the laying of a new tramway. The curve west of Alipašin Most, opened in 1901, provided a direct connection (primarily) for freight traffic between the Brod main line and the Narentabahn; note also the balloon loop linking the new station with the Ostbahn.

Courtesy Österreichisches Staatsarchiv, Vienna

From the earliest days of the Great War the railways of Bosnia-Hercegovina were heavily involved in the transportation of the sick and wounded to the various hospitals around the province. On 2 August 1914 the three Büssing buses operated between Banja Luka and Jajce by the k.u.k. Militärbahn Banjaluka–Doberlin were withdrawn and sent to Sarajevo to be used as ambulances; in May 1915 they were reported to be in very poor condition. By the beginning of 1915 the BHLB had fitted out three permanent ambulance trains, each capable of carrying 60 stretcher and 120 sitting cases; two more would be put into service before the end of hostilities. Additionally, in 1914–15 rakes of older four-wheeled coaches were stationed on strategic lines (the Krivajatalbahn, the k.u.k. Militärbahn Simin Han–Bulatovci and the Ostbahn) for use during the various campaigns against Serbia. Above, on 9 November 1914 wounded are taken off an ambulance train standing on the tracks of the Sarajevo tramway on the Appel Quay near the Latin Bridge where in June 1914 Gavrilo Princip had fired the two bullets which a month later brought Europe to war. The train had been hauled from the interchange sidings at Sarajevo station by tram locos nos. 2 and 11 (*Berislav Sekelj collection*). Below, after the successful assault on Mt. Lovćen and subsequent conquest of Montenegro by Austro-Hungarian forces in January 1916 lightly wounded men wait to be loaded onto an ambulance train at the BHLB's southern terminus of Zelenika (*courtesy Österreichisches Staatsarchiv, Vienna*).

Klose 0-6+4 no. 323(?) and an unidentified compound 0-8-2 halt at Uskoplje. The youthful Archduke Maximilian of Austria is believed to have been on board the train, travelling to the Imperial naval harbour at Kotor where he served as the captain of a corvette.

Courtesy Österreichisches Staatsarchiv, Vienna

Following the Central Power occupation of Serbia in the autumn of 1915, Austria-Hungary began work on a connecting line from the Ostbahn terminus of Vardište to the SDŽ 760m gauge system at Užice. As the railway would have to overcome the lofty Šargan mountain the challenge was formidable. Several routes were surveyed but, with considerations of cost and speed of construction to the fore, that chosen involved steep gradients and zig-zags. Plans for the new line included a significant expansion of the facilities at Vardište, among which was the provision of a semi-roundhouse for six locomotives, which would presumably be employed on banking duties. The entire project was abandoned in September 1916 after Romania's declaration of war on the Monarchy. When the railway was finally built by the SHS in 1924, both curves and gradients were generous, obviating the need for banking locos. Vardište remained a small passing station.

Courtesy Österreichisches Staatsarchiv, Vienna

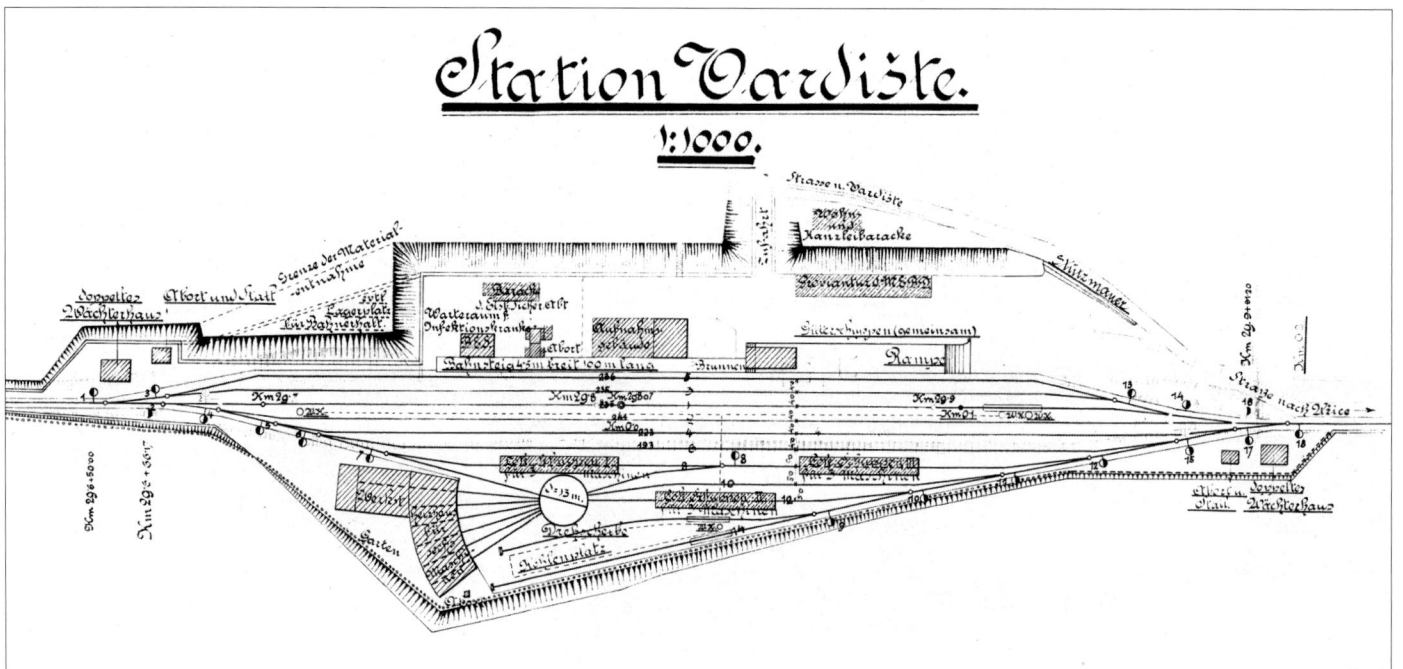

In Bosnia-Hercegovina nothing better illustrated the futility of war than the k.u.k. Militärbahn Simin Han–Bulatovci. Construction began in September 1914 to provide a supply route for Austria-Hungary's invasion and anticipated occupation of Serbia. But instead of easy victory, the Monarchy suffered humiliating defeats in its three campaigns against the small Slav state in the autumn and winter of 1914. The uneasy lull in the fighting that followed meant that by the time the 34.3km-long military railway was completed at the end of April 1915, there was little traffic for it. However, it was upgraded over the summer for the planned Central Power attack on Serbia and was extremely busy for 4–5 weeks in the autumn of 1915. With Serbia quickly overwhelmed, operations were suspended in January 1916 and the railway, which had cost over four million kronen, had been lifted by January 1917. During its construction, the army hired surplus 'half-duplex' 0-4-0Ts from the BHLB. In the upper picture, an unidentified member of the class shunts the exchange sidings at Simin Han whilst no. 26 (KrLi 1438/1883) perches on one of the many wooden viaducts so characteristic of this steeply graded line through the hills of eastern Bosnia.

Both: courtesy Österreichisches Staatsarchiv, Vienna

II : ROYALIST YUGOSLAVIA

The back end of Slavonski Brod station in the mid-1930s showing the narrow gauge platforms in the middle distance. Within weeks of the outbreak of the Great War, it became evident that the demands of wartime traffic were overwhelming the transhipment facilities between the BHLB and the MÁV at Bosanski Brod. To create a second, albeit smaller, interchange, in 1915 a third rail was laid between the standard gauge track between Bosanski and Slavonski Brod, finally providing a narrow gauge link to the latter. This, the Hungarians insisted, was for military purposes only. Thus, for example, they rejected out of hand the application of the Kopp company to lay at its own expense a narrow gauge siding into its flourmill and warehouses at Slavonski Brod, which would have enabled it to ship its products directly to Bosnia-Hercegovina; instead, it had to continue sending its goods half a kilometre in a 1435mm gauge wagon to Slavonski Brod, where they had to be transhipped to the 760mm gauge. Even in war, the integrity of the antipathies between Budapest and Vienna had to be maintained. After 1918 the SHS made Slavonski Brod the northern terminal of the 760mm gauge line to Sarajevo. Narrow gauge services into Slavonski Brod finally ceased in the early 1960s.

Keith Chester collection

The location and date of this photograph are unknown but believed to be somewhere on the Dalmatiner-Bahn in the 1920s. The slight dip in the track suggests a reason for this apparently abrupt derailment, which could have ended up as something far more calamitous. The three locos are Klose 0-6+4s: the rear train machine one of the 1900–01 batch (JDŽ class 185) whilst the derailed engine and that on the right are the 'strengthened' version of 1901 (JDŽ class 186). Note too the small oil tank for supplementary oil firing next to the cab on the upturned locomotive.

Herbert Moser collection

The 2-4-0Ts of class IIa 3 were remarkably camera shy, a shame as they were rather appealing locomotives. The Krauss factory at Linz supplied three in 1885 for the opening of the Mostar–Metković line, and shortly afterwards its Munich works erected four for the Doboj–Simin Han branch, which commenced operations in April 1886. All seven survived to be renumbered by the JDŽ in 1933, by which time most had gravitated to Doboj MPD, principally for use on the nearby Karanovac–Gračanica branch. The final two were withdrawn in the 1950s. No. 176-003 was KrLi 1266/1885.

Courtesy Železnički Muzej, Belgrade

In July 1923 rack and adhesion 0-6+4 no. 728 (WLF 2149/1913) and a sister loco pause at Pale on what was presumably a test run following overhaul at Sarajevo works. The Ostbahn was frequently used for such purposes though it was of course not possible to try out the newly repaired rack equipment on this adhesion only section of railway.

Keith Chester collection

Grand Hotel Petka. Gruž (Gravosa).

At Dubrovnik a 3.285km-long 760mm gauge tramway was opened in December 1910 to provide a connection with the BHLB station and the centre of town. It proved an immediate success and two additional tramcars had to be purchased in 1912 to augment the initial fleet of five (all from Weitzer of Graz). Above, the green which had originally adorned these tramcars had given way to blue by the time this commercial postcard was produced in the early 1920s; this distinctive livery, so redolent of the Adriatic, would be replaced by one of yellow round about 1930. Below, the town terminus was just opposite the 15th-century Pile Gate and comprised a short loop, just visible in this view from 1941.

Both: Alfred Moser collection

Over the winter of 1915–16 the railways of Bosnia-Hercegovina were hit by an acute motive power shortage, one that would persist until the end of the war. The response was to draft in locomotives from other parts of the Monarchy. However, in the general absence of powerful tender locomotives on the narrow gauge lines of either Austria or Hungary, the gap had to be somehow filled by low to medium powered tank engines, primarily kkStB U class 0-6-2Ts and the MÁV 490 class 0-8-0Ts. The very swiftness of the collapse of the Central Powers in the Balkans in the autumn of 1918 meant that many of these locos remained on the territory of the newly created Kingdom of Serbs, Croats and Slovenes, and were taken into SHS stock or allocated to industry. After working on the Steinbeisbahn in the war 0-6-2T no. U.36 (KrLi 5799/1905) spent some of the interwar years at Ljubija mine, where this photograph was taken (*Slobodan Rosić collection*). 0-8-0T no. 490,015 (Bp 2861/1912) was in Croatia at the end of hostilities, was taken into SHS stock and renumbered 81-001 by the JDŽ in 1933. A number of these very useful engines was always to be found in Bosnia-Hercegovina (*courtesy Železnički Muzej, Belgrade*).

In the spring of 1915 the Serb government, with the aid of Entente credits, ordered locomotives and rolling stock for the SDŽ in America, including 12 760mm gauge 2-8-0s from Alco. These were delivered within a few months but, following the occupation of Serbia by the Central Powers that autumn, were taken into kukHB stock. From March 1916 onwards some were transferred to Bosnia-Hercegovina to ease motive power shortages in the province. In April 1918, as kukHB no. 4103, the future no. 84-001 (Alco 55097/1915) was allocated to Sarajevo where, basking in the sun, it posed for the official photographer in the mid-1930s. For the next five decades up to half-a-dozen of these engines were always to be found in Bosnia-Hercegovina.

No. 83-030 (Bp 4993/1929) was one of 44 0-8-2s the Budapest works delivered as reparations to Yugoslavia in 1929. These locomotives were intended principally for trains on the new 760mm gauge main line from Belgrade to Sarajevo, which was opened to through traffic on 30 October 1928. Reparations played a not insignificant rôle in the reconstruction of Yugoslavia's railways after the Great War. A document dating from the early 1930s records that since 1920 the country had received the following 760mm gauge locomotives and rolling stock from Germany and Hungary: 84 locomotives, 120 passenger carriages, 40 postal and luggage vans, and 4,326 freight wagons from the former, and 79 locomotives and 430 wagons from the latter.

Both: courtesy Železnički Muzej, Belgrade

The election of the Stojadinović government in 1935, which coincided with the belated and slow recovery of the Yugoslav economy after the Great Slump of 1929, provided a long awaited opportunity to modernise the country's narrow gauge railways. In the late 1930s main lines were upgraded to permit a radical overhaul of the services offered to passengers, at the heart of which lay seven Ganz designed three-car diesel-mechanical sets. The Budapest firm supplied all mechanical parts and the power bogies, whilst the bodywork and final assembly were completed at Slavonski Brod. Painted in a striking silver-grey livery, they were put into service in July 1938.

Both: courtesy Železnički Muzej, Belgrade

Nova čelična elektr. varena kola kolosek 0.76 m.
izradena 1939 god. u radionici Maribor.

Another component of the JDŽ's modernisation programme was the provision of new passenger rolling stock for its 760mm gauge railways, the first since the influx of German reparations in the early 1920s. These all-steel vehicles were designed at Sarajevo and an order for 40 was placed in 1940. Owing to pressure of work in the central repair shops, 15 III class carriages were sub-contracted to Maribor and Smederevo. A further twenty were completed later that year at Slavonski Brod and fifty more were planned to follow from Sarajevo in 1941. Sadly, the outbreak of war prevented this.

Courtesy Železnički Muzej, Belgrade

In the autumn of 1916 a crude 17km-long 600mm gauge field railway was hastily laid between Bijeljina, centre of the fertile Posavina region, and Bosanska Rača on the river Sava to permit the transportation of foodstuffs to other parts of Bosnia-Hercegovina. It was upgraded for locomotive working in 1917, and extended by about 20km to the coal mines at Ugljevik in the summer of 1918. The line was rebuilt in stages to 760mm gauge in 1922–25. Even after reconstruction, the facilities for the transhipment of coal and cereals, both bulk goods, to and from the river barges at Bosanska Rača remained basic.

Keith Chester collection

In the 1920s and 1930s the physically isolated Bosanska Rača–Ugljevik branch fell under the aegis of first Direkcija Belgrade and then Direkcija Subotica, both of which employed on the line four or five of the 0-6-0Ts O&K had supplied in 1922 as reparations. One of these engines forms the backdrop for this group portrait of most (if not all) of the staff of Bijeljina station in the late 1920s; predictably, the railway workers have conspired to stand in front of the last two digits of the number plate, preventing the identification of the loco.

Keith Chester collection

The HDŽ, similar to railway administrations elsewhere in Nazi-occupied Europe, provided trains for the deportation of Jews from Sarajevo, first to the transit camp at Krivica near Travnik, and subsequently to the Jasenovac death camp in Croatia. Other transportations of people were more fortunate. This photograph, simply titled 'Kolonizacija Bijeljina 1942', is believed to show the relocation of the German populations from several villages near Bijeljina to ethnically more suitable areas closer to the heart of the Reich.

Courtesy Historijski Muzej Bosne i Hercegovine,

After the Axis invasion of Yugoslavia in April 1941 Rome annexed territory around the Bay of Kotor thereby placing the final sixteen kilometres of the former Dalmatiner-Bahn from Igalo to Zelenika under Italian sovereignty. This section of railway was duly absorbed into the FS. The 760mm gauge railways in the other areas of Yugoslavia occupied by Italy were nominally under the control of the HDŽ though operated on a daily basis by the Italian army in the shape of 2ª Sezione Militare di Esercizio. Two class 85-hauled passenger trains cross somewhere in Dalmatia in 1941–42. Italian soldiers are packed into the open wagons behind each locomotive to offer some protection against Partisan sabotage attacks, all too frequent on these lonely stretches of railway.

Ivo Angelini

The narrow gauge lines of Bosnia-Hercegovina were favoured targets for the Partisans during the Second World War and there were numerous 'diverzije'. The date and location of this particular one are unknown, but 0-8-2 no. 83-076 (KrLi 6035/1909) and several carriages and wagons have been tossed off the track. Railwaymen and soldiers pause briefly for their photo to be taken before starting the practiced task of getting the train back on the rails and repairing the damage. The sabotage looks more dramatic than it really was: the disruption would be a wearisome inconvenience at most, and trains would soon be running again. Today, no. 83-076 is preserved in Austria.

Courtesy Železnički Muzej, Belgrade

29

During the Second World War multi-layered conflicts of intense brutality were fought over what had been the Kingdom of Yugoslavia: a war of liberation against the Axis occupiers, a civil war among the nationalities of Yugoslavia, and an ideological struggle between the Communists and Royalists for the future of the country. Top, in the Drina gorge men of the Ustaše militia swarm along the tracks of the Ostbahn. Middle, shifting alliances led to many cases of collaboration. Italian soldiers and Chetniks pose together at Jablanica during Operation Weiss, launched in January 1943 to encircle and eliminate Tito's forces. The resulting Battle of the Neretva (March 1943), however, proved a great victory for the Partisans. Bottom, an official send off for Wehrmacht officers at Bijeljina.

All: courtesy Historijski Muzej Bosne i Hercegovine, Sarajevo

30

Between the late summer of 1941 and the end of hostilities, the Partisans under Tito's leadership operated their own trains in liberated territories in many parts of Yugoslavia. The longest (both in length and duration) of these Partisan railways in Bosnia-Hercegovina was that centred on Jajce where Tito set up his headquarters in August 1943. Trains were run south first to Donji Vakuf, and later extended to Bugojno and Gornji Vakuf. In the other direction, in November services were restored to Drvar, as well as on other short sections of the former Steinbeisbahn. Axis forces retook Jajce in January 1944 and Drvar the following May, bringing to an end the Partisan-operated Railway of the National Liberation Army (ŽNOV). 0-6+4 no. 185-036 (Arad 103/1901) stands at Bugojno in November 1943. ŽNOV performed a crucial rôle in the transportation of Partisan fighters and supplies to the front, as well as foodstuffs etc. for the civilian population.

All: courtesy Historijski Muzej Bosne i Hercegovine, Sarajevo

III : THE TITO ERA

In the early 1950s a tram and trailer head down Maršala Tita street towards the Baščaršija. Behind, Tito's name boldly proclaims the new Socialist world on the wall of the National Bank above the eternal flame monument for the victims of fascism.

Keith Chester collection

To win support for what was an abrupt break with the past, Tito's activists covered Yugoslavia with propaganda. There is barely a picture of the construction of the two standard gauge Youth Railways in Bosnia-Hercegovina which does not show a bridge, a wall or a skip daubed with a slogan of some sort or, as here, 'written' in stones on the bare earth. In 1946–47 Tito was pressing strongly for the union of Yugoslavia and Albania, but Enver Hoxha was increasingly hostile to this and in 1948 sided with Stalin in the dispute between Belgrade and Moscow.

Courtesy Železnički Muzej, Belgrade

Though the central repair shops in Sarajevo suffered some bomb damage, they were able to continue working throughout the war. For the 1945 May Day celebrations in Sarajevo, held just three weeks after the city's liberation, no fewer than 13 newly repaired locomotives were taken from the works over the tracks of the tramway to Marijin dvor to be displayed as symbols of Bosnia-Hercegovina's renaissance. Also on public view was one loco deemed beyond repair, bearing the inscription 'This is what the occupiers left us.' It must have been in a very bad condition: this class 83 bearing the markings of the HDŽ was returned to service after the war.

Courtesy Železnički Muzej, Belgrade

The German retreat from Sarajevo northwards to Brod in April 1945 was accompanied by one final orgy of destruction in Bosnia-Hercegovina, during which the old k.k. Bosna-Bahn main line suffered sorely. Its speedy rehabilitation was a priority for Tito's régime, with huge numbers of men and women engaged on rebuilding it. The Brod–Zenica section was reopened in July 1945, further south to Sarajevo a month later. The three photographs show work on reconstructing bridges over the river Bosna between Doboj and Zenica. The loco is a class 82 0-8-0T (possibly no. 82-005), fitted with an auxiliary tender and adorned with red stars.

All: courtesy Historijski Muzej Bosne i Hercegovine, Sarajevo

The Communists came to power in Yugoslavia in 1945 with the intention of not just reconstructing their shattered country but also modernising it. In Bosnia-Hercegovina this meant regauging some existing 760mm gauge railways and building new standard gauge ones. The first of the latter was the 92km-long Brčko–Banovići Youth Railway, completed in 190 days in 1946. That summer the wagons of Klose 0-6+4 no. 185-028 (KrLi 4489/1901) are being offloaded somewhere on the final section between Živinice and Banovići. With its 6.5t axle loading this relatively powerful compound was ideal motive power for the construction trains.

Courtesy Železnički Muzej, Belgrade

The Croat Zkonko Springer worked on the Živinice–Banovići section of the Brčko–Banovići Youth Railway in the summer of 1946.

Early July 1946 I took a train from Osijek via Vinkovci to Slavonski Brod where I had to change to a 760 mm narrow gauge railway… we travelled to Doboj first and then changed to a train for Tuzla leaving it at Lukavac. Army trucks were waiting for us there and drove [us]… to a small village Gornje Zivinice [and then]… on a rather narrow road along a small gauged railway line following a rivulet until one reached a narrow valley encircled by thick forest... Colleagues jumped onto a grassy floor grumbling and swearing and with some astonishment viewed the camp that would be their 'home' for 2 months now on.

Next morning we were waken up at 6 AM and after a formless face washing went to fetch a brown liquid alias for 'coffee' and a chunk of white bread for breakfast. The morning roll call and lifting the flag cum singing of the anthem was at 7 AM after that Camp's commander read the daily orders. At about 8 AM some 100 mostly male comrades left the camp to pick up the tools and proceeded to first working plot. The tools on disposal were shovels, pickaxes and heavy hoes as well as dozens of wheelbarrows. The first task was to clear undergrowth and bushes along a stretch that marked by wooden profiles and posts for the future railway line. Groups of tens formed of which appointed comrades were in charge and started working… The first working day finished at 4 PM after 8 hours of backbreaking earthwork… When the clearing work ended we started a cutting for the railway line. The excavated material had to be carted away few hundred meters to form a fill for the rail bedding… [A] new tool appeared at the site. It was large wooden stump with two handlebars fixed at each side to be used for compacting. The lifting and dropping of this heavy stump became one of the dullest jobs but allowed intervals of relaxation from sometimes...

One day in August we were called back from work and boarded trucks that took us to nearest station of the forest railway called 'shemandufer' (from French 'chemin de fer'). A line of flat wagons were waiting for us pulled by a puffing locomotive with a huge conical shaped chimney topped up with a wide mesh to prevent sparking. We got axes and shovels, boarded wagons when the train started ascended slowly a track of 600 mm gauge [sic] leading towards the slopes of Konjuh Mountain where the forest was on fire.

Zkonko and his comrades then set to putting out the fire. The 'shemandufer' was part of the former 'Victoria' 760mm gauge forestry railway, opened by Moritz Lisska in 1907. The full story of Zkonko's time on the Brčko–Banovići railway, as well as on the Šamac–Sarajevo Youth Railway in 1947, can be found on http://www.cosy.sbg.ac.at/~zzspri/index.html

Lašva 1947. A standard gauge railway running through the Bosna valley between Šamac and Sarajevo had first been mooted in 1878–79. It was proposed again in the mid-1890s but politics delayed its authorisation until 1912–13. Work on the project was just beginning when war put a halt to it. The scheme was revived after 1945 and the first train from Šamac steamed into Sarajevo on 16 November 1947. 242km in length, this Youth Railway was built in a record 230 days by large numbers of young people who worked enthusiastically (at least when the cameras were around) in return for little more than free food, clothing and rudimentary accommodation. Youthful enthusiasm of itself could not, however, construct a main line railway. Much of the detailed planning had been completed in 1913–14 and the project was in the hands of a large team of trained engineers and technical experts (as well as political commissars).

Left and above: courtesy Železnički Muzej, Belgrade, and below: courtesy Historijski Muzej Bosne i Hercegovine, Sarajevo

Following the opening of the Šamac–Sarajevo Youth Railway, the 106km-long narrow gauge line north of Zenica to Doboj closed to all traffic. The short Doboj–Usora section (for trains to Teslić and Pribinić) was realigned to allow access to Doboj Novi, located some distance from the original 760mm gauge station. In 1951 the former kkBB main line immediately south of Doboj is slowly returning to nature and will soon be lifted.

Courtesy Historijski Muzej Bosne i Hercegovine, Sarajevo

However, between Sarajevo and Zenica the 760mm gauge track was retained, primarily to avoid the inconvenience and expense of transhipping between the two gauges the iron ore imported through the harbours of Dubrovnik and Ploče and destined for Zenica steelworks. The two lines ran in parallel, weaving in and out of each other.

Courtesy Železnički Muzej, Belgrade

Narrow and standard gauge tracks cross at the southern entrance to the junction station of Lašva.

Courtesy Železnički Muzej, Belgrade

In the spring of 1966 2-6-2 no. 73-006 (KrLi 5965/1908) stands at Bosanski Brod with a local train to Doboj. The Allies had bombed both Bosanski and Slavonski Brod heavily towards the end of the Second World War with the oil refinery at the former and the engineering plant at the latter major targets. In the days before 'precision' bombing many other buildings, such as Hans Niemeczek's once magnificent neo-Orientalist station at Bosanski Brod, were also hit.

John Millbank

The symbolism of the destruction was perhaps appropriate: following the 1947 opening of the standard gauge Šamac–Sarajevo Youth Railway Bosanski Brod was no longer served by main line trains, just local services to Pribinić. These were cut back to Teslić in 1954 and Derventa in 1968. No. 185-028 (KrLi 4489/1901) pulls away from Bosanski Brod in August 1961. A wave of narrow gauge closures across Bosnia-Hercegovina on 1 June 1969 finally put Bosanski Brod station out of its misery.

Michael Schumann

A Doboj Novi–Sarajevo local passenger, hauled by 4-8-0 no. 11-053 (Bp 7412/1955), easily overtakes narrow gauge 0-8-2 no. 83-166 (Bp 5679/1948) on 9 July 1966.

Peter Gray

UNRRA 0-8-0 no. 19 (HKP 8065/1945) shunts a short train of coal at Zenica on 9 September 1961. With the opening of the standard gauge Sarajevo–Ploče line in November 1966 the narrow gauge between Sarajevo and Zenica finally closed. Local passenger services on the 760mm gauge north of Sarajevo, which continued for a while after 1947, had petered out by the late 1950s.

Paul Booth

An unidentified class 92 2-6-6-0 steams through the Bosna valley at the head of a lengthy freight in 1948. These powerful Henschel Mallets were built in 1917–18 primarily for the heavy ore trains emanating from Bor copper mines in the German sector of occupied Serbia. The initial batch of 20 (of which 19 were taken into SHS stock) was augmented by a second one of 30 in 1922. From the 1920s many of these fine locomotives were allocated to Bosnia-Hercegovina where they put in much hard work for some decades. The allocation of 28 in 1939 had been halved to 14 by 1950 and over the next couple of years they disappeared for good from the rails of ŽTP Sarajevo.

Slobodan Rosić collection

The determined effort made at the end of the 1930s to provide enhanced levels of comfort and shorter journey times for the passengers of Yugoslavia's 760mm gauge system proved short lived. The outbreak of war in March 1941 saw the seven three-car Ganz diesel railcars placed in store at Sarajevo, mainly due to the lack of diesel fuel, and one was subsequently damaged beyond repair in a bombing raid. Upon the return of peace the remaining six sets were quickly put back into service, given appropriate names and painted a good Socialist red. Within a few years they had lost their names and had been returned to their original silver-grey livery. This photograph of *Sutjeska* in the Neretva valley can thus be dated to the late 1940s.

Courtesy Železnički Muzej, Belgrade

Socialist Realism
à la Bosnie

A booking office clerk...

..the šef stanica at Drvar (after 1945 an employee of the JDŽ)...

...the driver of no. 97-015 (WLF 1467/1901) reads his train orders...

...a proud fireman.

All: courtesy Železnički Muzej, Belgrade

As was the fashion of the time children's railways were built in major cities in Yugoslavia in the late 1940s. Most of these were 600mm gauge but that at Ljubljana was 760mm. Its sole locomotive was Klose 2-4-2 no. 178-003 (KrLi 3155/1895), which stands at the head of a rake of freshly painted stock in a station still far from complete. Five decades previously, this loco had been hauling the fastest passenger trains on the BHStB.
Courtesy Železnički Muzej, Belgrade

When the Red Army swept through Hungary in 1944, no. 83-040 (Bp 5003/1929) was one of six 0-8-2s under repair at Debrecen works. Unlike the others, it was taken back to the USSR as war booty and set to work on the Khar'kiv children's railway. Now numbered LK 83-1, it was used there until scrapped in the early 1950s.
Dmitry Olentsevich collection

In the late 1940s the Yugoslavs ordered ten class 83s with large bogie tenders from MÁVAG as reparations. Six had been taken into JDŽ stock when the breach between Tito and Stalin ended further deliveries. The remaining four languished in the Budapest works until 1950–51 when, rebuilt to metre gauge, two apiece were put to work at the Ózd and Diósgyőr steelworks. No. 485,5002 (Bp 5683/1950) was photographed at the latter working a passenger train between Diósgyőr and Pereces on 7 September 1964.
Peter Gray

At the end of 1949 the central repair shops, now named the Vaso Miskin-Crni enterprise (after an eponymous Partisan hero), was selected as one of the first places to trial the Yugoslav experiment with self-management, with the initial workers' council being elected in March 1950. A few years later, a class 73 2-6-2 undergoes repair inside the main workshops.
Courtesy Historijski Muzej Bosne i Hercegovine, Sarajevo

The postwar explosion of Sarajevo's population overwhelmed the town's veteran 760mm gauge tramway. The catalyst for change was the availability of redundant standard gauge PCC tramcars from the Washington DC system, and in 1956–57 the decision was taken to regauge the tramway. Above, in a scene very typical of the narrow gauge system in its last years, motor car no. 67, one of 18 built in Sarajevo in 1947–54, and trailer no. 19 were photographed in the loop at Baščaršija heading for Sarajevo Novo station. Built by O&K in 1903 for the steam tramway at Bohumín, the trailer was acquired by the electric Kleinbahn Pirano–Portorose in 1916, and came to Sarajevo when that closed in 1953 *(Stan Keyse / Andy Carey collection)*. Top right, on Sunday, 9 October 1960, the last day of narrow gauge operations, the 'new' secondhand 1435mm gauge cars were exhibited at Marijin dvor together with some old 760mm ones. The crowds seem more interested in ex-Washington tramcar no. 29 (a St. Louis Car product) than in no. 69. Originally supplied as a trailer in 1909, it was rebuilt as a motor car in 1930 and ran as no. 65 for twenty years before being renumbered 69 in 1950. At the time of its withdrawal it was the last tramcar running in Sarajevo with a clerestory roof *(Keith Chester collection)*.

2-8-2 no. 85-001 (Bp 5030/1930) waits for the off at Mostar with an express bound for Sarajevo on 10 September 1964. These handsome machines, of which a total of 45 were eventually placed in service pre-war, were the third stage of 35 years of narrow gauge express passenger engine development for Bosnia-Hercegovina, following on from the Klose compound 2-4-2s of 1894 and superheated simple 2-6-2s of 1907. They may also be considered as the swansong of locomotive design for the 760mm gauge state railways: post-1945 accessions to stock consisted of either further construction of pre-Great War types (e.g. class 83), 'off-the-shelf' products (e.g. UNRRA 0-8-0s), or locos for industry (e.g. Czechoslovak 0-6-0Ts or 0-10-0s).

Alan Sprod

No. 83-127 (KrLi 5071/1903) takes water at Hum in September 1964 whilst no. 83-136 (KrLi 5474/1906) and an unidentified class 2-8-2 prepare to depart on the last leg of a Sarajevo–Dubrovnik express. In latter years, the compound 0-8-2s were concentrated on the lines of the former Dalmatiner-Bahn. Double heading of the heavier trains, both passenger and freight, was common.

Blake Paterson

On a blazing hot day in July 1969, nos. 83-065 (Jung 3547/1923) and 83-120 (KrLi 7500/1919) head a short train from Bileća to Trebinje past lake Trebinjsko near Arslanagića Most. Hydro-electric schemes along the river Trebišnjica in the 1960s, which would create artificial lakes near both Trebinje and Bileća, required the complete realignment of the Hum–Bileća railway between Aleksina Međa and Miruše in 1962.

Both: John Gulliver

'Gone fishin' ... '

Courtesy Železnički Muzej, Belgrade

No. 83-156 (ĐĐ 51/1948) at Drežnica, September 1964.
Blake Paterson

Electrification of the Sarajevo–Mostar–Metković main line was first suggested in 1910 and would have involved the erection of an electricity works at Jablanica for power supply. When the idea of the construction of a standard gauge railway from Sarajevo to Ploče to replace the old narrow gauge was revived in the late 1940s, this too was to be electrified, again utilising a new hydro-electric power plant at Jablanica. Lack of money meant that this was the only part of the scheme to immediately proceed for there was also the need to provide electricity for the rapidly expanding conurbation of Sarajevo. It became one of the largest projects undertaken in Bosnia-Hercegovina in the early 1950s. In the event the standard gauge Sarajevo–Ploče railway did not open until November 1966, and then only with diesel traction, electrification would take another three years. The construction site at Jablanica was served by its own 760mm gauge system and had its own loco, ex-JDŽ no. 189-016 (KrLi 2691/1892). The damming of the river Neretva led to the flooding of a section of the original Narentabahn and a new 760mm gauge line built to a standard gauge profile between Jablanica and Konjic was opened in 1954. The propaganda slogans above the loco say (left) 'Long live Tito's Five-Year Plan' and (right) 'We will provide 21,000 wagons of gravel for the dam'.

Courtesy Historijski Muzej Bosne i Hercegovine, Sarajevo

Train control on the single-tracked Bosnian narrow gauge remained fairly rudimentary to the end with operations dependent for their smooth running on either the telegraph or telephone. Yet, even on intensively trafficked sections such as the Ivan pass bottleneck, it was a system that worked very well thanks to the dedication of the railwaymen of the JDŽ and JŽ.

Courtesy Železnički Muzej, Belgrade

Ploče: big ship, narrow rails. After decades of politicking and fruitless discussion the development of the port of Ploče was finally authorised in 1936. The next year work was begun on a 21km-long railway between Metković and Ploče, which was finally opened on 25 November 1942 during the Italian occupation. However, the incomplete port was blown up in the immediate aftermath of the Italian capitulation the following September. Both the line and port were reopened to traffic on 25 July 1945 and played an important rôle in the reconstruction of Bosnia-Hercegovina in the 1950s and 1960s.

Courtesy Železnički Muzej,
Belgrade

Ploče: before the fall. Work on the long-planned standard gauge line between Sarajevo and Ploče began seriously in 1963 following the securing of a $35 million loan from the World Bank. Surveying of the new railway is just beginning at Ploče where ships and a class 83(?) can be glimpsed in the background.

Courtesy Železnički Muzej,
Belgrade

High above, the new standard gauge Sarajevo–
Ploče railway takes shape whilst, below, we are
in a world in which seemingly little has changed
in a century as a short narrow gauge passenger
train passes a horse and cart.

Courtesy Železnički Muzej, Belgrade

Trains on the adhesion sections of the
Sarajevo–Mostar–Metković railway were at first
primarily in the hands of radial tanks, but from
1900–01 onwards examples of the enlarged
tender version, which subsequently became
JDŽ class 185, began to appear in the Neretva
valley. Though the line was upgraded after
the Great War to permit the use of the heavier
0-8-2s, the Klose 0-6+4s continued to find
employment on lighter duties until it was closed
in November 1966. Shortly before this the
official photographer turned up at Konjic MPD
and recorded an unidentified member of the
class balanced on the shed's turntable.

Courtesy Železnički Muzej, Belgrade

Between 1894 and 1919 the Floridsdorf works in Vienna delivered a total of 38 rack and adhesion 0-6+4s to Bosnia-Hercegovina, BHStB class IIIc 5 and JDŽ class 97. Within a few years of their introduction, efforts were being made to find a replacement for them, notably the two 0-4-6-0R semi-Mallets of 1906 and the electrification proposal of 1910. For three decades from 1912 onwards various alternative routes to the Ivan pass were considered, but for various reasons they mostly fell by the wayside, leaving the class 97s to carry the burden of hauling and banking trains up and down the Ivan pass until 1966 and the Komar pass until 1975. An unidentified class 97 receives attention inside Konjic MPD.

Courtesy Železnički Muzej,

Less than two months before the fires are dropped for the final time, 0-6+4R no. 97-033 (WLF 2254/1915) stands at Konjic surrounded by concrete sleepers and the new rails for the standard gauge Sarajevo–Ploče railway, rapidly approaching completion. 22 September 1966.

Michael Schumann collection

The loco crews at Konjic gather in front of another unidentified 0-6+4R, probably for a valedictory photograph taken shortly before the end of narrow gauge rack working over the Ivan pass on 5 November 1966.

Courtesy Železnički Muzej, Belgrade

The mid to late 1960s witnessed many closures on the JŽ 760mm gauge lines both in Bosnia-Hercegovina and Serbia. Nevertheless, the retention and modernisation of parts of the once extensive narrow gauge system was foreseen and, to this end, in 1967–68 Đuro Đaković supplied 12 four-car diesel railcar sets (class 802), followed by 40 B'B' diesel hydraulics (class 740) in 1968–71. Yet within a few years ŽTP Sarajevo, mindful of its new image as the operator of modern standard gauge railways, suddenly reversed its position. By 1979 all the remaining 760mm gauge lines in Bosnia-Hercegovina had been closed, rendering a largish fleet of new and expensive diesel locomotives and railcars redundant almost overnight. A few were sold but most were quietly scrapped. This all lay in the future, however, when this class 802 two-car set departed from Čapljina for Dubrovnik.

Courtesy Železnički Muzej, Belgrade

On 5 November 1966 no. 85-041 (SlBr 28/1940) sets off from Hadžići into the night with the last through Ploče–Sarajevo narrow gauge train on the final adhesion section between Bradina and the Bosnian capital. This working marked the end of the old 760mm gauge Narentabahn and large crowds gathered along the line to bid their farewell. The two makeshift headboards fitted to the loco at Hadžići caught the mood. The upper one reads 'Ćiro: Good luck for your retirement. The railwaymen's collective at Hadžići.' That below says simply 'Thank you dear Ćiro. The inhabitants of Hadžići.' 'Ćiro' was the affectionate nickname given to their narrow gauge railway by the peoples of Bosnia-Hercegovina: the origins and precise meaning of the word are obscure. Charlie Lewis, who visited Bosnia-Hercegovina both before and after the closure of this line, wrote that with its disappearance 'the heart had gone out of the narrow gauge'. It is a conclusion hard to disagree with.

Slobodan Rosić collection / Borba Foto Arhiv

During the 1960s the Dalmatian coast, with Dubrovnik at its heart, was promoted as a destination for mass tourism. As the visitors, increasingly car-bound, flooded into Dubrovnik, so its tramway came to be seen as an anachronism. With its equipment life expired, it was easier to close than modernise. At the beginning of 1969 notice was given that the system would close in early 1971. This was not to be: on 7 March 1970 as tramcar no. 5 descended to Pile, its brakes failed. Out of control, it overshot the tracks and ended up in the dry moat beneath the old city walls, causing three fatalities and about 30 injuries. A temporary service was then put into place running between the station and the hospital, but even this was ended on 20 March. Above, tramcar no. 5 climbs the steep hill away from the Pile Gate terminus in 1959 and, left, rests on its side following the accident on 7 March 1970 which sealed the fate of the tramway.

Both: Alfred Moser collection

LAST DAY AT LAPAD

Nick Lera takes up the story. 'I arrived by railcar from Čapljina at Dubrovnik station in the early evening of Thursday 19 March. Outside stood an immaculate vintage tram. "Surely this should be in a museum," I said (in German) to the elderly conductor. "Yes, that's where it's going tomorrow," he replied. "So are there new trams coming?" I asked. "Oh no, we're closing down at mid-day," he said, "and then buses will take over! There'll be free rides in the morning and we shut at 12.00." In probably the most serendipitous moment ever in my gricing career I had arrived unbeknownst on the eve of Last Tram Day! So I took the museum piece to my Lapad hotel for the night. Next day, Friday 20th, I was sorely vexed as to how to apportion my precious 16mm film, rail or tram, but the trams won the day and I filmed less steam than I wanted as a result. Normal revenue operations ran until 10a.m. (so 20 March just about qualifies as the last service day) but thereafter travel was gratis. The packed trams were a delight with all the townsfolk enjoying their free rides. Promptly at noon the service was halted, all nine vehicles of the system formed a straggled procession and one by one filed into the depot. The gates shut at 12.30 and that was that.'

16mm frames by Nick Lera; map by J. C. Gilham, courtesy LRTA

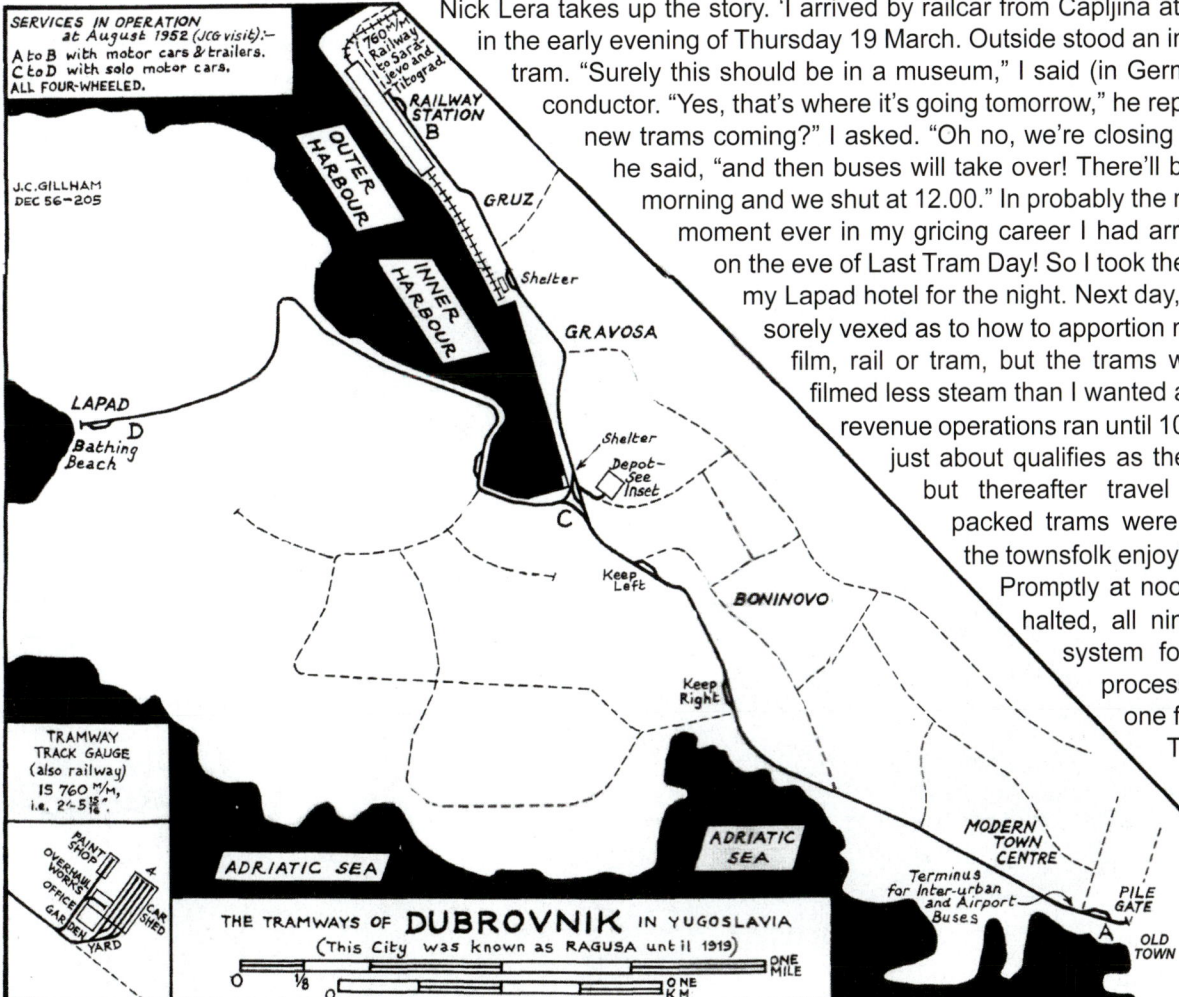

SERVICES IN OPERATION at August 1952 (JCG visit):-
A to B with motor cars & trailers.
C to D with solo motor cars.
ALL FOUR-WHEELED.

J.C.GILLHAM DEC 56-205

760 M/M Railway to Sarajevo and Titograd

RAILWAY STATION B

OUTER HARBOUR

INNER HARBOUR

GRUZ

Shelter

GRAVOSA

LAPAD D
Bathing Beach

Shelter
Depot-
See Inset
C

Keep Left

BONINOVO

Keep Right

TRAMWAY TRACK GAUGE (also railway) IS 760 M/M, i.e. 2'-5½".

PAINT SHOP
OVERHAUL WORKS
OFFICE
GARDEN
CAR SHED
YARD

ADRIATIC SEA

ADRIATIC SEA

MODERN TOWN CENTRE

Terminus for Inter-urban and Airport Buses

PILE GATE
A
OLD TOWN

THE TRAMWAYS OF **DUBROVNIK** IN YUGOSLAVIA
(This City was known as RAGUSA until 1919)

0   1/8   ONE MILE
0   ONE KM

0-6+2T no. 189-025 (KrLi 2872/1893) stands at the head of a short freight at Donji Vakuf on 24 August 1961. Had the Spalato-Bahn ever been built (see page 8), Donji Vakuf would have become a junction of considerable importance. Instead, politics within the Austro-Hungarian Monarchy and then the complete redrawing of borders after 1918 condemned it to remaining a small country station which sprang into life several times a day but otherwise slumbered.

Martin Smith

0-6+4 no. 185-025 has its smokebox cleaned at Donji Vakuf in August 1964. 45 of this more powerful and partially enlarged variant of the Klose radial tank were built in 1900–01 for the opening of the Dalmatiner-Bahn. This particular example came from Arad, works number 102 of 1900; under the terms of the 1897 Ausgleich between Vienna and Budapest one third of all locomotives and rolling stock supplied to Bosnia-Hercegovina had to be erected in the Hungarian half of the Monarchy.

Michael Schumann

Having panted its way up the 45% (1 in 22.22) rack section from Goleš, supershine no. 97-009 (WLF 1336/1900) catches its breath at Komar before plunging into the 1,361m-long summit tunnel and beginning the descent to Donji Vakuf. 24 August 1961.
Martin Smith

Winter grips Komar station as an unidentified class 97 waits with a short passenger train bound for Goleš and Lašva. The 'S.Z.' board (Svršetak Zupčanice) indicates the end of the rack section.
Courtesy Železnički Muzej, Belgrade

Sarajevo, July 1966. Whilst 2-8-0 no. 84-002 (Alco 55098/1915) indulges in some vigorous shunting, no. 84-004 (Alco 55100/1915) prepares to meet, if not its maker, then the ovens of the steelworks at Zenica.

Between the terminus of the 1947 Youth Railway from Šamac and the old 760mm gauge station in Sarajevo lay a distance of about one kilometre. To make interchange easier for passengers, the narrow gauge was extended to Sarajevo Novo in 1954. The facilities were minimal (three tracks and little more) and pictures invariably show passengers milling between the trains. 0-6-2T no. 72-011 (Hens 8580/1907) was one of several ex-SDŽ locos stationed at Sarajevo in the 1960s for shunting and empty stock duties.

All: Peter Gray

The first proposal for constructing a superheated version of the compound 0-8-2 introduced in 1903 came in October 1907 (see page 7). This drew on the experience garnered with the type III b 5 2-6-2s which had first gone into service that year and, above all, the marked superiority of the superheated NÖLB 0-8+4s (class Mh) of 1906 over the compound variant (class Mv) of 1907. It took, however, almost another two years before the first superheated 0-8-2 went into service on the BHLB, but it would prove to be one of the most successful 750–760mm gauge locomotives ever designed. Superheated no. 83-095, seen here departing Međeđa on 29 August 1974, was manufactured by Krauss, Linz in 1914, works number 6893.

Robert Stevens

The 0-8-2s were associated with the Ostbahn from its opening in 1906 till its closure in 1978. An unidentified member of the class shunts at Međeđa in the 1950s.

Courtesy Železnički Muzej,
Belgrade

With the opening of the Ostbahn in the summer of 1906 came a number of schemes, initially using motorised road transport, to improve communications along the Drina valley south of Ustiprača, in particular to the two larger communities of Goražde and Foča. The first proposal for a railway seems to have been made by the Bahnindustrie AG of Budapest, which in the autumn of 1911 sought the concession for either a forestry railway or a common carrier local railway between Ustiprača and Foča where there were considerable stands of timber. Nothing came of this. A short forestry railway was eventually opened from Brod na Drini, just south of Foča, in 1932 and three years later the JDŽ began the construction of a 760mm gauge main line from Ustiprača to Bileća. This quickly ran into difficulties and the first 42km from Ustiprača to Foča was only completed in September 1939. After the war, there was no interest in pursuing this scheme and the putative trunk route slipped into the torpor of a rural branch line. Above, no. 83-066 (Jung 3548/1923) heads a short passenger train near Foča on 30 August 1974 (*Robert Stevens*) whilst, below, no. 83-016 (Bp 4978/1929) pauses at Goražde with a mixed in 1966 (*John Middleton*).

The familiarity of the narrow gauge: no. 83-066 (Jung 3548/1923) runs through the Drina gorge near Međeđa on 29 August 1974.

Robert Stevens

On a glorious autumn's day in October 1972 an unidentified class 85 heads a freight through the Drina gorge.

Peter Lemmey

Between 1928 and 1968 the Ostbahn finally found an appropriate rôle for itself as part of the 760mm gauge main line connecting Belgrade and Sarajevo. Over its rails ran lengthy expresses between the two capitals. East of Višegrad trains were routinely double-headed for the ascent of the Šargan mountain, the famous 'Šargan 8', which today has been revived as a museum railway. On 10 July 1966 nos. 83-011 (Bp 4973/1929) and 85-037 (SlBr 24/1940) brew up at Višegrad ready for the stiff climb ahead.

Peter Gray

Boys will be boys. Shortly before the closure of the Ostbahn on 28 May 1978 no. 740-106 (ĐĐ 891/1971) enters Međeđa with a train packed with delegates for a preparatory meeting for the XI Congress of the Communist Party of Yugoslavia. The loco was subsequently one of four of these now redundant but almost brand-new locomotives sold to Banovići coal mines for line working. In 1998 it was purchased for spare parts by the Steiermärkische Landesbahnen in Austria.

Slobodan Rosić collection / Borba Foto Arhiv

IV : FORESTRY RAILWAYS

In her book *Bosnisches Skizzenbuch* published in 1900 Milena Preindlsberger-Mravozić wrote that Dobrljin had 'in recent years grown from a village to almost a town due to the sawmill' erected there in 1893–94. The original gauge of the forestry railways laid by the Otto Steinbeis concern in the 1890s was 800mm. These were at first horse worked, but locomotives were introduced on the Vučja poljana line in 1898. The internal system at the Dobrljin sawmill, also 800mm gauge, was locomotive worked from the beginning, principally by an 0-6-0T which emerged from Krauss' Munich factory in 1892, works number 2645.

Courtesy Steinbeis Archiv

Within a decade the entire Steinbeis operation centred on Dobrljin had been dwarfed by a bold new scheme. To exploit vast additional concessions Steinbeis erected a huge sawmill at Drvar and built a 102km-long railway from Oštrelj to Knin. The choice of 760mm gauge for this railway, the fact that it was engineered to quasi main line standards, and its strategic geo-political location, strongly suggest that Steinbeis' backers in Sarajevo and Vienna anticipated that this railway, known as the Steinbeisbahn, would eventually be incorporated into the wider system of narrow gauge railways in Bosnia-Hercegovina.

Courtesy Steinbeis Archiv

Bosnia-Hercegovina with its lightly laid, sharply curved and steeply graded lines would appear to have been ideal Mallet country. But the state railways pointedly ignored the type: the master Gölsdorf sitting in Vienna was a steadfast opponent of this form of articulation. The privately owned Steinbeisbahn, however, did not hesitate. O&K 892/1901, illustrated here, was the first of 13 Mallets of various wheel arrangements Steinbeis acquired from the Berlin firm and Maffei between 1901 and 1909. Today it is preserved at Banja Luka.

Courtesy Freunde der Eisenbahn, Hamburg

On a bleak day in the winter of 1906 an O&K 0-4-4-2T has overturned at Oštrelj, revealing its private parts to the world.

Keith Chester collection

On the high plateau through which much of the Steinbeisbahn ran, heavy snowfall was an annual curse faced by its operators and unsurprisingly the railway owned a number of snowploughs. Much of the removal work though was done by large teams of men equipped with shovels.

Courtesy Steinbeis Archiv

62

Road no. 2, O&K 892/1901, pauses near the water station at km 10.

Courtesy Steinbeis Archiv

A pair of Maffei 0-4-4-0Ts pose for their portrait in the loop at Tiskovec shortly after the opening of the Knin–Drvar main line of the Steinbeisbahn in March 1902. The locos are separated by a wagon to distribute weight over the many wooden viaducts on this stretch of railway.

Courtesy Steinbeis Archiv

No. 18 (Maff 2996/1909) works a heavy train of logs and sawn timber to Knin. This was the second of the three powerful 0-6-6-0Ts Maffei supplied to the Steinbeisbahn between 1907 and 1909.

Courtesy Steinbeis Archiv

The two-cylinder compound 0-10-0Ts Maffei designed for the Steinbeisbahn were perhaps the most impressive of all the locomotives constructed for the narrow gauge railways in Bosnia-Hercegovina. They exuded brute force, and kept their promise. Five were delivered to Drvar between 1911 and 1914. The last two, nos. 25 and 26, were slightly modified, driving on the third rather than the fourth axle though the experiment was not repeated. Beginning in 1912 Maffei began to supply a superheated simple version to the BHAG at Teslić. It seems that in all a total of seven was ordered for Teslić, but it only received three: of the remaining four, two were sent in 1915 direct to the k.u.k. Militärbahn Čardak–Jajce, operated by the Steinbeisbahn, and the last two were never completed during the war, being sold off to private railways in the 1920s. No. 26 (Maff 3888/1914) heads a short passenger train at Drvar in the interwar years.

Frank Steuber collection

This 3hp motorised draisine was produced in the Drvar workshops of the Steinbeis enterprise just before the outbreak of the Great War.

Courtesy Steinbeis Archiv

In August 1919 the new Kingdom of Serbs, Croats and Slovenes stripped Otto Steinbeis of all his remaining property in Bosnia-Hercegovina and formed a state-run company called ŠIPAD to take over the sawmills etc. at Drvar and Dobrljin, as well as the associated railways. A six-wheel tank wagon bearing the markings of its new owners undergoes repair inside the well equipped shops at Drvar in the 1920s. The size of the works hall bears testimony to the scale of the entire Steinbeis operation in northwest Bosnia.

Courtesy Steinbeis Archiv

No. 32 was the last of three 0-10-0s ŠIPAD ordered from O&K in 1924 (works number 10830). These 350hp locos were the only new motive power acquired for the Steinbeisbahn between the wars: the railway was a drain on resources and ŠIPAD was disinclined to invest in it.

Keith Chester collection

Somewhere along the Steinbeisbahn

Deep in the forest between Oštrelj and Srnetica...

...an isolated loading point and the small community that grew up around it...

...an O&K 0-10-0 heads a short train of stone in the 1920s...

...a bogie carriage and a motley collection of four-wheel coaches bring up the rear of a mixed hauled by a Maffei 0-4-4-0T.

All: courtesy Steinbeis Archiv

With the Axis invasion of Yugoslavia in April 1941, Italy was finally able to realise its long-held irredentist claims to large parts of the Dalmatian coast. Railways in the occupied zone, including the southern section of the former Steinbeisbahn, were operated by the 2ª Sezione Militare di Esercizio, an officer of which was photographed in front of Maffei 0-10-0T no. 26 at Knin.

Ivo Angelini

This poor quality photograph is supposed to show the destruction at Drvar after the German air attack at the end of May 1944, from which Tito just escaped with his life. On the other hand, included in the line up of locomotives are two class 92 2-6-6-0s. It is known that the JDŽ allocated three of these large Mallets to Drvar in 1946 so it is possible that the photograph is incorrectly dated. Or is it evidence that the HDŽ transferred some of these Henschel 2-6-6-0s to the Steinbeisbahn during the war?

Courtesy Historijski Muzej Bosne i Hercegovine, Sarajevo

The three Mallets were intended principally for banking heavy trains to the reversing point at Oštrica, a duty performed with little distinction, and they spent most of their time at Drvar in store until removed in 1949. The three gentlemen, a little overdressed for a depot yard, are perhaps conducting an inspection to decide their fate.

Keith Chester collection

To alleviate the acute shortage of motive power on the former Steinbeisbahn after the Second World War the JDŽ placed an order for six 0-10-0s with the Škoda factory at Plzeň on 17 June 1947. As with many locomotives constructed in the early years after the war, the design was not entirely new. In this particular case it had evolved out of one for a five-coupled tank locomotive being worked on by German engineers in the latter years of the war for the 785mm gauge Upper Silesian system. This was a development of the six 0-10-0Ts Chrzanów had supplied in 1929 as class Tw9 and would form the basis for the PKP's postwar 0-10-0Ts for Upper Silesia, classes Tw47 and Tw53. Final design work and manufacture of Škoda type 60Lo1 were carried out at Plzeň under the aegis of engineer Otto Lederer, who is standing in the middle of the group of men in front of no. 1935 (Škoda 1935/1949), one of three 0-10-0s dispatched to Yugoslavia on 29 March 1949. As this date suggests, not every country immediately stepped into line with the economic embargo imposed upon Yugoslavia in 1948 at the USSR's behest.

Michal Štrublík collection

UNRRA no. 34 (HKP 8080/1946) heads a Prijedor–Srnetica passenger train near Bravsko in August 1968.

Michael Schumann

Klose radial tank no. 189-022 (KrLi 2857/1893) at Srnetica on 8 July 1966. Following the Second World War the JDŽ assumed responsibility for the 'main lines' of the old Steinbeisbahn and began transferring some sorely needed motive power, including a small number of the 'radialka' tanks. For all the intricacies of the Klose mechanism, these locomotives were very successful in Bosnia-Hercegovina and were distinguished by their longevity of service: the last one disappeared in 1967, the Summer of Love.

Peter Gray

The final generation of steam motive power on the Steinbeisbahn. Lesley Folkard's 1966 photograph of no. 1933 (Škoda 1933/1949) and no. 34 (HKP 8080/1946) standing side-by-side at Srnetica MPD illustrates perfectly the sheer bulk of the Škoda 0-10-0s. These worked heavy freights and the Porter 0-8-0s the lighter passenger trains between Prijedor and Drvar until 1970–71 when they were replaced by the new Đuro Đaković class 740 B'B' diesel hydraulics. The modernisation was short lived: the former Steinbeisbahn was closed in stages between 1975 and 1978.

Subsequent to the withdrawal of Lenin's Russia from the Great War in 1917, Austria-Hungary's armies fighting on the Eastern Front were moved westwards, followed by all their equipment. By the spring and summer of 1918 considerable quantities of surplus locomotives, rolling stock and permanent way materials were arriving on the Italian and Balkan fronts, albeit too late to influence the outcome of the war. 0-8-0T no. 36 (Maff 3786/1912) was transferred to the Steinbeisbahn from the Krechowice forestry railway in Galicia in May 1918. With the coming of peace it remained on the books of ŠIPAD and after 1945 was taken into JDŽ stock. The loco spent much of the 1960s employed as station pilot at Jajce, where it was photographed in August 1962.

Michael Schumann

No. 18 (HKP 8066/1945) waits for the off at Šipovo on 8 July 1966. The 34 0-8-0s Porter supplied via UNRRA in 1945–46 helped avert a humanitarian disaster in Yugoslavia. When the Second World War ended, Split was the only port open for desperately needed food imports and until April 1946 the Steinbeisbahn was the only railway connecting Split with the interior of the country. The influx of these 0-8-0s was literally a lifesaver in the immediate aftermath of the war and, though some members were allocated elsewhere in Yugoslavia, the majority of the class remained active on the Steinbeisbahn until 1970–71.

Peter Gray

Few photographs better encapsulate the hold of the narrow gauge railways of Bosnia-Hercegovina on the imagination than this magnificent view of *Jablanica* (Maff 2731/1907) in a gorge in the Krivaja valley.

THE KRIVAJATALBAHN

Having successfully negotiated in 1898–99 the contract for the extensive timber stands in the Krivaja valley, the Austrian Louis Eissler and the German Richard Ortlieb set up a new company called the Bosnische Forstindustrie AG (confusingly Steinbeis called his 1900 company the Bosnische Forst-Industrie AG). Eissler & Ortlieb established their headquarters at Zavidovići, a passing station on the BHStB main line. From here a 118km-long forestry railway was built to supply logs to what soon became one of the largest and most modern sawmills in Europe. Within a few years Zavidovići had grown from a hamlet of less than a dozen dwellings to a sizable township complete with a company school and clinic, employee housing and a church.

This pannier tank (Mödl 25/1874) was presumably the first loco to run in the Krivaja valley where it was employed on construction trains. It was one of two 0-6-0PTs supplied in 1874 by F.X. Mannhart's engineering works at Mödling near Vienna to the 750mm gauge Montanbahn Rostoken–Marksdorf (in present-day Slovakia). It came to Bosnia in 1900 bearing the number 2, but was soon given the name *Zavidović*. In this evidently posed photograph from the late 1920s the engine was looking in fine fettle as it shunted amidst the stacks of sawn wood. Apart from the original seven Hügel & Sager 0-4-0Ts (see pp. 4–5), this veteran machine was the oldest loco to have worked in Bosnia-Hercegovina.

Unless otherwise indicated all photos in this section are courtesy Yvonne Illich, curator of the Illich Regenstreif Family Archives (IRFA) and Ortlieb Family Archive (OFA)

For line work Eissler & Ortlieb purchased five 0-6-2Ts from Krauss, Munich in 1900–01, a design which, with variations, found widespread use on forestry railways in Central Europe. To increase their range of operations in the lengthy Krivaja railway, auxiliary tenders were routinely attached to these engines, as with this unidentified example pausing at Krivaja station before the Great War.

Piccolo (KrMü 4391/1901) was the first of five 0-6-0Ts delivered by the Krauss Munich and Linz factories between 1901 and 1904. These small engines were intended for work on the numerous, often temporary, branches off the 'main line' of the Zavidovići–Kusače forestry railway, as well as on shunting turns.

It soon became apparent that something more powerful than the 0-6-2Ts was required for the fiercely graded Olovo–Kusače section of the Krivajatalbahn and for this purpose Eissler & Ortlieb acquired five 0-4-4-0Ts from Maffei between 1904 and 1910. These were a slightly modified version of the three Mallets delivered to the Steinbeisbahn in 1903. *Hofrat* was Maffei 2716 of 1908. The name (lit. 'court counsellor') remains even in modern, republican Austria an honorific title eagerly sought by civil servants.

That the Krivajatalbahn was eventually to evolve into a common carrier is evidenced by this steel viaduct across the river Krivaja in the lower reaches of the valley.

Courtesy Narodna Biblioteka Srbije, Belgrade, Album 'Krivaja : societé anonyme forestière : anc. Eissler & Ortlieb, 1930'

Most of the bridges and viaducts (some quite substantial) along the Krivajatalbahn were, however, constructed of wood, over one of which *Jablanica* (Maff 2731/1907) eases a train of logs in the 1920s. For the first decade of its existence, this 0-4-4-0T was called *Forstmeister* but like all the other Mallets on the railway was stripped of its very Germanic name in the postwar, independent South Slav state and given a more appropriate, local moniker.

Franz Xaver Komarek began manufacturing steam railmotors at his factory in Vienna in 1903. That year and the next he delivered six 760mm gauge examples to the NÖLB. In 1904 an order was placed for a saloon railmotor for the use of the directors and senior officials of the Bosnische Forstindustrie. No doubt high ranking visitors to the line also got to ride in it. The photograph shows the saloon (works number 22) shortly after its arrival at Zavidovići in the summer of 1905. Drive was on the leading single axle, giving this superheated compound a 2-2-4T wheel arrangement. Due to the nature of its duties, it probably saw even less service than most other steam railmotors (which was not much: those on the narrow gauge lines of the NÖLB were by and large out of use by 1911–12) and the whole affair was something of an indulgence by Messrs. Eissler & Ortlieb. Nevertheless, the Komarek, which was never given a name or number, was still on the books of the Krivaja company in 1934.

Keith Chester collection

Principal dimensions of Komarek 22/1905:	
Cylinders:	175mm x 250mm (hp)
	250mm x 250mm (lp)
Driving wheel diameter:	620mm
Trailing wheel diameter:	620mm
Total length:	10140mm
Weight, empty:	15t
Weight, in service:	17.7t
Heating surface:	17m²
Superheated surface:	2.25m²
Grate area:	0.63m²
Boiler pressure:	13atm
Water capacity:	1.5m³
Coal capacity:	0.6m³
Max. permitted speed:	45km/h

At the other end of the scale was this crude, wooden draisine for the track gangs. Others were not so fortunate. The Bosnische Forstindustrie was not permitted to operate passenger services over its forestry railway, and the five ex-kkBB four-wheel carriages the company owned in 1914 were, officially at least, restricted to the transportation of its employees only. However, it is hard to believe that in the remote Krivaja valley that a ride was refused to normal mortals.

Bosnische Forstindustrie

EISSLER & ORTLIEB

Bahnbetrieb

From its earliest days Eissler & Ortlieb operated an ambulance car, converted from a four-wheel III class carriage supplied by the Hernalser Waggonfabrik in Vienna to the kkBB in 1879, to transport sick and injured employees (the profession of lumberjack was a dangerous one) to the company's clinic at Zavidovići. By 1914 a further two, one of which appeared on a publicity postcard put out by O&K Budapest, had been constructed new for the railway. Primitive though the vehicle may seem, it was at least sprung and bumping along in it over the tracks of a forestry railway was still preferable to a long and agonising journey jolting over a muddy road in a horse-drawn cart.

Keith Chester collection

Orenstein és Koppel cég pusztaszentlőrinci gyárában készült sebesültszállító kocsi 760 mm. nyomtávban.

The Krivajatalbahn became, by default, one of the principal supply routes for the three ill-fated campaigns Austria-Hungary conducted against Serbia in the summer and autumn of 1914. To cope with the massive increase in traffic, the military authorities requisitioned large numbers of locomotives: from within Bosnia-Hercegovina itself came engines from forestry railways and the BHLB (mainly Klose 0-6+2Ts), and from the wider Monarchy kkStB U class 0-6-2Ts and MÁV 490 class 0-8-0Ts. Also sent to Zavidovići were nine lightweight 0-8-0Ts (MÁV class 492) and a Budapest works type 78 0-6-0T no. 3003 (Bp 2826/1914), all of which had been hastily ordered new for the military in late 1914 and early 1915. With operations winding down in the Krivaja valley five of the 0-8-0Ts and no. 3003 were transferred to the k.u.k. Militärbahn Simin Han–Bulatovci in the late spring of 1915 (see p. 21). This 0-6-0T was moved to Borsa in January 1916. Thereafter, it led a peripatetic existence, finally being withdrawn as MÁV no. 399,102 on the Nyírgyháza system in 1961.

Keith Chester collection

Eissler & Ortlieb's undertaking in the Krivaja valley was second only to that of Otto Steinbeis. The very size of this loading point underscores the scale of the operation.

Logs are manhandled off the four-wheel trucks of the forestry railway for processing in the sawmill at Zavidovići. Much of the sawn timber was eventually dispatched to the ports of Metković or Dubrovnik for export. In 1926 Zavidovići was the sixth busiest freight yard in all of Yugoslavia. It is thus all the more surprising that Eissler & Ortlieb was an unprofitable concern and had to be bailed out by the Yugoslav state in 1934.

Similar to Steinbeis at Drvar, the Bosnische Forstindustrie operated some electrified sidings within its large sawmill at Zavidovići to minimise the fire risk. In 1908 the company acquired an electric locomotive from Kolben & Co. of Prague. This, however, soon proved to be underpowered and within a few months Eissler & Ortlieb were looking for a more powerful replacement. The photograph shows a small electric loco of unknown provenance (unfortunately the large works plate cannot be read) shunting inside the Zavidovići factory between the wars.

The politics of the Krivajatalbahn were complicated. A forestry railway built basically to Lokalbahn standards by the provincial government of Bosnia-Hercegovina, it was leased for 30 years to the privately owned Bosnische Forstindustrie AG; after this period, the railway was to become a state-operated common carrier. For Sarajevo, this was a rather neat arrangement: the annual rent paid by Eissler & Ortlieb would easily cover the construction costs, eventually providing the cash-strapped province with a railway for free; it also kept the feuding interference between Vienna and Budapest at arm's length. In 1921 Eissler & Ortlieb's contract was renewed for 20 years but the company was in financial trouble and fiercely disputed the government's insistence it operate public trains. The way out of the impasse was for the Ministry of Forests and Mines to have four locos built at state expense to work these trains. The choice of motive power was bold: 0-10-0Ts that were powerful in appearance and performance. For four decades these impressive looking machines (a fifth one followed in 1930) worked the daily passenger and other heavy trains up and down the Krivajatalbahn, where they were nicknamed 'Bimba'. The five 0-10-0Ts were, like all previous Eissler & Ortlieb locos, given names, ranging from the mundane (e.g. *Pilana* or sawmill) to the worthy, such as the first constructed, *Industrija* (KrLi 1362/1925), seen here standing on the turntable at Zavidovići in the late 1920s.

An unidentified 0-10-0T hauls a long train of logs in the lower part of the Krivaja valley.

Post-1945, all five 0-10-0Ts were taken into JDŽ stock and the photographic evidence suggests they also lost their names. In August 1962 a very well cared for no. 1363 (KrLi 1363/1925) catches its breath (and some water) before tackling the next part of the 40% (1 in 25) climb between Olovo and Han Pijesak. The loco was previously named *Trgovina*.

Both: Michael Schumann

OTHER FORESTRY RAILWAYS

In 1913 Fadilpasič & Co. were granted a concession for a horse-worked forestry railway feeding their sawmill at Begov Han. Four years later due to the shortage of horses and feed in Bosnia-Hercegovina, the successor company, the Industrie- und Handels-Aktien-Gesellschaft 'Bosna', decided to rebuild it as a locomotive-worked line, with the first four kilometres awaiting inspection in July 1918. By this stage, a loco, believed to be 0-4-0T KrLi 2411/1890 purchased from the Usoratalbahn, was already on the railway. Thus, by the skin of its teeth, the 'Bosna' company opened the last loco-worked forestry railway in Habsburg Bosnia-Hercegovina.

By 1925 the line had expanded to 19.2km in length, all save 0.9km loco-worked. The company owned three engines, 35 wagons and six primitive Lowrys. Apart from the small Krauss 0-4-0T, the locomotives are unidentified. What these bare statistics do not reveal is that this relatively short and obscure forestry railway was one of the most spectacular in Bosnia-Hercegovina. These photographs, presumably taken in the early 1920s, show a forestry railway that with its zig-zags and trestles was almost American in appearance. Appropriately, the legendary Bosnian Shay (Lima 2504/1912) is said to have worked out its days here.

All: courtesy Narodna Biblioteka Srbije, Belgrade,
Album 'Industrijalno i trgovačko d. d. "Bosna" –
Begov Han / Atelje Đ. Božić'

This exquisitely proportioned 0-6-2T, essentially a slimmed down version of the kkStB class U, was one of nine erected by Krauss, Linz between 1897 and 1904. Five went to the Gurktalbahn in Austria, where they were designated class T ('T' for Treibach–Althofen), and four to Bosnia-Hercegovina. Three of the latter were acquired by the Bosnische Holzverwertungs-AG for its wood distillation plant at Teslić, but in a period of financial difficulty were hired out to the neighbouring Usoratalbahn until April 1904. The photograph was taken at Pribinić.

Dejan Škrebić collection

No. 1 (KrLi 3581/1897) was the first of the T type 0-6-2Ts, as well as being the first loco ordered by the BHAG for its Teslić operation. Pictured here on the premises of the distillation plant in the interwar years, this once elegant locomotive has been disfigured by the fitting of extended side water tanks. Today it is preserved at Teslić.

Courtesy Narodna Biblioteka Srbije, Belgrade, Album 'Destilacija Teslić'

Even though it is in a deplorable external condition (with almost as much rust on display as paint), KrLi 4657/1901 has retained its original appearance and is even still carrying its worksplate. Purchased by the Bosnian Bureau on behalf of the state-operated Usoratalbahn, this was the third T class 0-6-2T (the other two were KrLi 3838/1898 and 5091/1904) delivered to Bosnia-Hercegovina. It remained in the Usora valley until after the Second World War, but ended up as no. 691 at PDI Ilijaš Bosna, Srednje, where it was photographed in July 1964.

Michael Schumann

The third sawmill after Drvar (1901) and Zavidovići (1908) to employ electric traction for safety reasons was at Turbe where in 1913 the Ugar Forstindustrie AG laid a short internal line, electrified at 500v DC. Supplied by Brown, Boveri & Co. in Vienna (works number 837 of 1913) with mechanical parts by StEG, this 22hp loco could haul a load of 26t at 15km/h on the maximum grade of 25% (1 in 40).

Alfred Moser collection

0-4-0T Henschel 14526/1916 was a vestige of the Great War: sent by the Germans to Bor copper mines in Serbia, it was taken into SHS stock after 1918 and became JDŽ no. 172-005 in 1933. Subsequently sold into industry, it spent most of the post-1945 years working at the Impregnacija at Vitez. August 1962.

Michael Schumann

No. 1121-21 was one of approximately ten Ryś type 0-4-0Ts supplied to Yugoslavia in 1947–48 by Chrzanów. The number scheme was unusual (the numbers of most industrial locos simply repeated their works numbers), but no doubt reflected the fact that the majority, if not all, of these engines were first allocated to units of the Yugoslav National Army. Like no. 172-005, it too was photographed at Vitez in August 1962.

Michael Schumann

The first forestry railways in Bosnia-Hercegovina were short, horse-worked affairs in the late 1880s. The Usoratalbahn, opened in 1890, initiated the era of 760mm gauge forestry railways utilising steam locomotives, a concept that would flourish for the next seven decades until superseded by the lorry in the 1960s. But whether horse or locomotive-worked, all forestry railways relied on a variation of the four-wheel truck.

Slobodan Rosić collection

0-6-0T *Kuza* (Hens 20033/1924) takes water on PDI Vrbas in the 1950s. The first section of what developed into an extensive system of forestry railways running south of Banja Luka was hurriedly opened in 1918 primarily to serve the coal mine at Maslovare: by this stage of the Great War coal shortages throughout the Monarchy were extremely acute. The Vrbanje forestry railway was one of the many forestry railways in Bosnia-Hercegovina that were rarely photographed.

Slobodan Rosić collection

The last classic forestry railway in Bosnia-Hercegovina was the Maglić system, which finally closed in the autumn of 1978. Henschel 0-8-0T (works number 19126 of 1923) hauls a train of logs towards Miljevina in the 1960s. Known as *Teja*, this loco was the sole motive power on the Partisan railway operated by the Fočanska Republika between January and May 1942, reason enough to secure its preservation. The loco is currently plinthed at Dobrun.

Courtesy Železnički Muzej, Belgrade

After forestry the most important industrial sector in Bosnia-Hercegovina was mining, mainly for coal and iron. Mines ranged considerably in size, from small, often unprofitable, private concerns to large and lucrative state-owned ones. The short-lived coal mine at Lesljani near Dobrljin, above left, very much belonged to the former category *(Keith Chester collection)*. The first coal mine to electrify its internal and underground 500mm gauge lines was Breza in 1913–14. However, the most significant industrial electrification in the Habsburg era was at Vareš iron mine, where the 3.4km-long 550mm gauge line from the ironworks to Droškovac was so equipped in 1913 and employed two Ganz and two SSW locomotives by 1918. Above right, an unidentified electric loco emerges from an adit on 19 September 1955.

Courtesy Historijski Muzej Bosne i Hercegovine, Sarajevo

The ironworks at Zenica was established in 1892 but, despite ready access to coal and the kkBB to transport its finished products, the concern remained unprofitable until the Podlugovi–Vareš branch was opened in 1895. With abundant iron ore from Vareš now available at competitive rates, Zenica began to flourish and was transformed into the Eisenindustrie AG, Zenica in 1899. Recapitalised, the works were expanded and modernised in the early 1900s, with annual steel production tripling between 1900 and 1912. On 28 May 1908 three BHStB six-wheel open wagons await loading in the surprisingly tidy yard at Zenica. Note the piles of rails, which formed a considerable proportion of Zenica's output.

Keith Chester collection

In a Stalinist dash to create a new industrial base, the Communists rapidly expanded the coal mining and metallurgy sectors in Bosnia-Hercegovina after the Second World War. One beneficiary of this policy was the iron mine at Vareš where an extensive complex of 900mm gauge lines was laid out. This was a new gauge for Bosnia-Hercegovina, the Yugoslavs having adopted the German 'standard' narrow gauge for heavy industries.

Courtesy Železnički Muzej, Belgrade

The question of reparations to Yugoslavia after both wars was complex. As constituent parts of the Kingdom of Yugoslavia had fought on both sides in the Great War, reparations were made to 'Serbia', which is why many engines ran with SDŽ (and not SHS) plates in the 1920s. After 1945 Cold War politics were responsible for yet another complicated situation. Payments could not be made to a country deemed within the Soviet bloc but because, after 1948, Western powers entertained hopes of wooing Yugoslavia away from the Socialist camp, Germany provided Tito's régime with generous financial aid with which locomotives could be 'purchased'. Using this diplomatic sleight of hand Jung supplied 36 900mm gauge 0-4-0Ts to Vareš in 1953, four of which await track upon which to run.

Slobodan Rosić collection

The completion of the 66.7km-long Doboj–Tuzla–Simin Han branch in April 1886 opened up a region rich in natural resources. In 1900 Milena Preindlsberger-Mravozić hailed Tuzla as the 'Bosnian Manchester' and before the civil war in the 1990s the town was one of the wealthiest in Yugoslavia. A salt works was opened at Simin Han in 1884–85, and an even larger one at Donji Tuzla in 1890–91. The photograph shows 0-4-0T O&K 11284/1926 shunting at the latter on 10 September 1953.

Courtesy Historijski Muzej Bosne i Hercegovine, Sarajevo

The Kreka coal field near Tuzla was one of the largest in Bosnia-Hercegovina with the first mine sunk in 1885. Between the wars the colliery at Bukinje was one of the most productive. No. 1 (D.R.B. = Državni Rudnik Bukinje = State Mine Bukinje), photographed in the 1930s, is believed to be one of the two 0-4-0Ts erected at Slavonski Brod in 1924 using parts supplied by Krupp (works numbers 898 and 899), making these two small engines the first locos 'built' in independent Yugoslavia. Narrow gauge operations ceased at Bukinje following the rebuilding of the Doboj–Tuzla line to standard gauge in 1947–51. Today, its small loco depot and repair shops are home to some of the last active steam engines in the world.

Courtesy Železnički Muzej, Belgrade

The Yugoslav economy was hit very hard by the Crash of 1929 and it was only in the late 1930s, under the Stojadinović government, that investment again started to be made in infrastructure and heavy industry. One manifestation of this was Jugočelik (Yugoslav Steel), a grouping of steel works, iron ore and coal mines in northern Bosnia established as a conglomerate large enough to benefit from economies of scale and to compete in international markets. In 1940 Jugočelik acquired this 0-8-0T from Jung (works number 9385). After the war it was to be found at the Ljubija iron mines (where presumably it had been since 1940), but ended its working life on the Maglić forestry railway. Today it is preserved at Brod na Drini.

Courtesy Freunde der Eisenbahn, Hamburg

By 1940 no fewer than nine industrial railways fanned out in various directions from Živinice, serving mining and forestry interests in the area. One line, opened in 1928, ran to Bukinje, providing a physical connection with the state railway system and, more importantly for the local community, a passenger service. Apart from the forestry railways, these lines were worked by locomotives owned by the Kroacija D.D., one of which, sadly unidentified, is seen here on 31 May 1938.

Petar Savić collection

No. 81-007 (Bp 4277/1916) stands at Foča with a short passenger train which, since the locomotive was allocated to the nearby coal mine at Miljevina, was presumably for workmen. This dates this photograph to the 1950s as the 14km-long Foča–Miljevina line, to all intents and purposes a forestry railway also used for coal trains, was only taken over by the JŽ and upgraded to a common carrier in 1962. Before coming to Miljevina, this 0-8-0T was one of the many engines that worked at Banovići in the early postwar years.

Courtesy Železnički Muzej, Belgrade

600mm has always been a comparatively unusual gauge in Bosnia-Hercegovina, and one confined mostly to mining operations. The last active example is at Kakanj, where coal is brought 6.5km from the pit head at Haljinići to the screens on the standard gauge railway. Opened in the 1950s this short line was part of a larger, electrified 600mm system. Destruction was heavy in the civil war and when it was resurrected in the late 1990s, the Italian government provided as aid four 4w diesels built by Valente of Milan. On 9 March 2009, no. 2 (works number 1017 of 1997) hauls a long train of empty tubs back to the mine. The concrete masts of the former electrification are clearly visible.

Keith Chester

Banovići coal mines

Above left: Coal had been mined on a small scale in the Banovići area, with some of the mines even served by railways before the Second World War. Production was massively expanded in the immediate postwar years and the huge opencast mine became a prestige project for the new Communist régime. A standard gauge railway from Brčko was constructed by youth volunteers in 1946 to carry coal to fuel-starved Belgrade and Serbia. The mines themselves had their own dedicated 760mm system. This small unidentified loco is highly decorated and proudly bears a 'Tito Coal Mines' headboard for some now unknown festive occasion.

Courtesy Železnički Muzej, Belgrade

Above right: In the haste to start operations locomotives were taken from where they could be found and the early years at Banovići were marked by the use of a veritable potpourri of narrow gauge engines. This picture, which is simply identified as 'Banovići 1946', shows a JDŽ class 176 2-4-0T backing onto a rake of loaded wagons. The use of these locos at Banovići is not documented and this was presumably a short-term loan.

Courtesy Historijski Muzej Bosne i Hercegovine, Sarajevo

Below: Wagons are loaded with brown coal whilst an unidentified side tank loco patiently waits.

Courtesy Železnički Muzej, Belgrade

In 1922 Yugoslavia acquired as reparations thirty 90hp 0-6-0Ts from O&K, of which 22 were taken into SHS stock and eight were attached to the Ministry of Forests and Mines, under whose auspices they worked for various state-owned mines and forestry railways. The Ministry probably had the better end of the deal as the state railways never seemed to know quite what to do with these small engines. However, after 1945 some of the JDŽ machines started to be hived off to industry, either on loan or sold off completely. Banovići was an early recipient of such locos with no fewer than 17 recorded as working there at different times in the late 1940s and early 1950s. They undoubtedly fulfilled an immediate need and their presence allowed the huge undertaking to get underway: no. 71-008 (tentatively identified as O&K 10165/1922) assists in the removal of overburden. But, in the long term they were underpowered for the rough and tumble of Banovići and were replaced as soon as possible.

Both: courtesy Železnički Muzej, Belgrade

In the late 1940s the Yugoslavs placed orders with MÁVAG for 40 of the 1942 modified variant of the well known type 70 0-8-0Ts (MÁV class 490). 37 of these reparation deliveries reached Yugoslavia before the dispute with Stalin put a halt to further construction. Many were destined for Banovići, where 13–14 (depending on source) worked there until the last two (nos. 55-98 and 55-99) were withdrawn in the late 1970s. Following the split in the Socialist bloc, the West quickly stepped in to fill the vacuum created by the trade embargo. The 'package' included locomotives, amongst which were 15 0-4-0STs from Porter, all of which went to Banovići. The photographs show two obviously brand-new locos. Above, no. U-9 (HKP 8246/1950) stands in front of the as yet incomplete workshops at Banovići, together with a class 71, an unidentified industrial loco, a class 185 tender and another Porter. Below, no. U-5 (HKP 8242/1950) waits for a wagon to be loaded with coal. The supershine appearance of the 0-4-0STs is unlikely to have lasted long.

Both: courtesy Železnički Muzej, Belgrade

In their quest for suitable motive power for Banovići the Yugoslavs also turned to Czechoslovakia, ordering ten burly 245hp 0-6-0Ts from ČKD in the winter of 1946–47. Details of the transaction are somewhat murky (as were the politics of the time) and construction was delayed. All ten were dispatched on 8 February 1949. 65 years later several of these still survive and one is regularly steamed for shunting at Oskova. No. 25-28 (ČKD 2528/1949), however, suffered the indignity of finishing its days as a stationary boiler at the Resod Guming factory in Tuzla. 9 April 1998.

Keith Chester

During the 1990s the workshops at Banovići rebuilt the two articulated diesel hydraulics obtained from Niš in 1987 into three four-wheel diesel shunters, making these the latest (but hopefully not the last) 760mm gauge locomotives in Bosnia-Hercegovina. On 8 May 1997, no. 720-002 was sporting a dull red livery; today, all three are painted in green with yellow lining.

Keith Chester

'My cup runneth over.' Between 1976 and 1980 Banovići colliery purchased seven class 83 0-8-2s from the JŽ. Enthusiasts who visited the mine in 1996–97 after the signing of the Dayton Peace Accords were amazed to find some still in service. They had owed their survival to shortages of diesel fuel during the civil war in the early 1990s. No. 83-159 (ĐĐ 54/1948) takes water at Oskova on 15 March 1997.

Keith Chester

Perhaps even more surprisingly, nearly two decades on an 83 is still steamed most days as the small class 720 diesels are not powerful enough to shunt the heavy coal trains over the unloading chutes at Oskova, a duty which fell to no. 83-159 on 11 March 2014.

Bill Jamieson

ABBREVIATIONS

The following abbreviations are used to indicate locomotive builders:

Alco	American Locomotive Company, Schenectady, USA
Arad	János Weitzer, Arad, Hungary
Bp	Magyar királyi Államvasutak Gépgyára Budapesten, Budapest, Hungary; after 1925 MÁVAG (Magyar Állami Vas-, Acél- és Gépgyárak)
ČKD	Českomoravská-Kolben-Daněk A Sp, Prague, Czechoslovakia
ĐĐ	Đuro Đaković, Industrija Lokomotiva, Slavonski Brod, Yugoslavia; formerly Prva Jugoslavenska Tvornica Vagona i Mostova
Ganz	Ganz és társa Villamossági-, Wagon- Gép-, és Hajógyár R.t., Budapest, Hungary
Hens	Henschel & Sohn, Cassel, Germany
HKP	H.K. Porter Locomotive Works, Pittsburgh, USA
Jung	Arnold Jung Lokomotivfabrik GmbH, Jungenthal, Germany
KrLi	Lokomotivfabrik Krauss & Comp, Linz, Austria
KrMü	Lokomotivfabrik Krauss & Comp, Munich, Germany
Lima	Lima Locomotive and Machine Company, Lima, Ohio, USA
Maff	J.A. Maffei AG, Munich, Germany
Mödl	Maschinen- Lokomotiv- und Wagen-Bauanstalt in Mödling, Austria
O&K	Orenstein & Koppel AG, Berlin-Drewitz, Germany
Škoda	Škoda-Werke, Plzeň, Czechoslovakia
SlBr	Prva Jugoslavenska Tvornica Vagona i Mostova, Slavonski Brod, Yugoslavia
SSW	Siemens-Schukertwerke, Vienna, Austria
StEG	Maschinen-Fabrik der priv. österreichischen-ungarischen Staats-Eisenbahn-Gesellschaft, Vienna, Austria
Tub	Les Ateliers Métallurgiques; Nivelles, Tubize and La Sambre, Belgium
WLF	Wiener Lokomotivfabrik, Wien-Floridsdorf, Austria

Other abbreviations used are as follows:

BHAG	*Bosnische Holzverwertungs-AG* : Bosnian Wood Utilisation Company, Teslić
BHLB	*Bosnisch-Herzegowinische Landesbahnen* : Provincial Railways of Bosnia-Hercegovina (1908–18)
BHStB	*Bosnisch-Herzegowinische Staatsbahn* : State Railway of Bosnia-Hercegovina (1885–95)
BHStB	*Bosnisch-Herzegowinische Staatsbahnen* : State Railways of Bosnia-Hercegovina (1895–1908)
CO	*Compagnie d'exploitation des chemins de fer Orientaux* : Oriental Railways Operating Company
FS	*Ferrovie dello Stato* : Italian State Railways
GVI	*Gazdasági Vasutak Igazgatosága* : Division for Economy Railways (Hungary)
HFB	*Heeresfeldbahn* : German Military Railways
HDŽ	*Hrvatske Državne Željeznice* : Croatian State Railways (1941–45)
JŽ	*Jugoslovenske Željeznice* : Yugoslav Railways (1953–2005)
JDŽ	*Jugoslovenske Državne Željeznice* : Yugoslav State Railways (1929–53)
kkBB	*kaiserlich-königliche Bosnabahn* : Imperial-Royal Bosna Railway (1879–95)
kkStB	*kaiserlich-königliche österreichische Staatsbahnen* : Imperial-Royal Austrian State Railways
kukHB	*kaiserlich und königliche Heeresbahn* : Imperial and Royal Military Railway
kukHB Süd	Imperial and Royal Military Railway South (Organisation responsible for the operation of railways in the Austro-Hungarian sector of occupied Serbia 1916–18)
MÁV	*Magyar királyi államvasutak* : Royal Hungarian State Railways (1869–1946); *Magyar Államvasutak* : Hungarian State Railways (since 1946)
MPD	Motive Power Depot
NÖLB	*Niederösterreichische Landesbahnen* : Lower Austrian Provincial Railways
PDI	*Preduzeće ze Drvna Industrija* : Timber Industries Enterprise
PKP	*Polskie Koleje Państwowe* : Polish State Railways
SDŽ	*Srpske Državne Željeznice* : Serbian State Railways (1889–1919 & 1941–44)
SHS	*Državne Željeznice Kraljevina Srba, Hrvata i Slovenaca* : State Railways of the Kingdom of the Serbs, Croats & Slovenes (1919–29)
ŠIPAD	*Šumsko Industrijsko Preduzeće a.d.* : Forestry Industry Enterprise Co. (formerly Otto Steinbeis A.G.)
UNRRA	United Nations Relief and Rehabilitation Administration
ŽNOV	*Železnica Narodnooslobodilačke Vojske* : Railway of the National Liberation Army
ŽTP	*Železničko Transportno Preduzeće* : Railway Transport Enterprise